U0223761

影印版说明

1.《传感材料与传感技术丛书》中第一个影印系列 MOMENTUM PRESS 的 *Chemical Sensors:Fundamentals of Sensing Materials & Comprehensive Sensor Technologies*（6 卷，影印为 10 册）2013 年出版后，受到了专家学者的一致好评。为了满足广大读者进一步的教学和科研需要，本次影印其 *Chemical Sensors:Simulation and Modeling* 系列（5 卷，每卷均分为上下册）。本书是第 5 卷 *Electrochemical Sensors* 的上册（第 5 卷 1～6 章内容）。

2. 原书的文前介绍、索引等内容在上下册中均完整呈现。

材料科学与工程图书工作室

联系电话 0451-86412421

0451-86414559

邮　　箱 yh_bj@aliyun.com

xuyaying81823@gmail.com

zhxh6414559@aliyun.com

CHEMICAL SENSORS
SIMULATION AND MODELING
Volume 5
Electrochemical Sensors

EDITED BY **GHENADII KOROTCENKOV**

影印版

化学传感器：仿真与建模

第5卷 电化学传感器

下 册

哈爾濱工業大學出版社
HITP HARBIN INSTITUTE OF TECHNOLOGY PRESS

黑版贸审字08-2014-077号

Ghenadii Korotcenkov
Chemical Sensors : Simulation and Modeling Volume 5 : Electrochemical Sensors
9781606505960

Originally published by Momentum Press, LLC
English reprint rights arranged with Momentum Press, LLC through McGraw-Hill Education (Asia)

图书在版编目（CIP）数据

化学传感器：仿真与建模. 第5卷，电化学传感器 = Chemical Sensors : simulation and Modeling. Volume 5 : Electrochemical Sensors. 下册：英文 /（摩尔）科瑞特森科韦（Korotcenkov, G.）主编. —影印本. —哈尔滨：哈尔滨工业大学出版社，2015.1
（传感材料与传感技术丛书）
ISBN 978-7-5603-4899-5

Ⅰ.①化… Ⅱ.①科… Ⅲ.①化学传感器-研究-英文 ②电化学-化学传感器-研究-英文 Ⅳ.①TP212.2

中国版本图书馆CIP数据核字（2014）第244957号

材料科学与工程
图书工作室

责任编辑 张秀华　许雅莹　杨　桦
出版发行 哈尔滨工业大学出版社
社　　址 哈尔滨市南岗区复华四道街10号　邮编 150006
传　　真 0451-86414749
网　　址 http://hitpress.hit.edu.cn
印　　刷 哈尔滨市石桥印务有限公司
开　　本 787mm×960mm　1/16　印张 11
版　　次 2015 年 1 月 第 1 版　2015 年 1 月第 1 次印刷
书　　号 ISBN 978-7-5603-4899-5
定　　价 60.00元

（如因印刷质量问题影响阅读，我社负责调换）

Contents

Preface ix

About the Editor xiii

Contributors xv

Part 3: Electrochemical Biosensors

7 Nanomaterial-Based Electrochemical Biosensors 251

N. Jaffrezic-Renault

1 Introduction 251

2 Nanomaterials: Fabrication, Chemical and Physical Properties 252

2.1 Conducting Nanomaterials 252

2.2 Nonconducting Nanomaterials: Magnetic Nanoparticles 254

3 Conception and Modeling of Amplification Effect in Nanomaterial-Based Enzyme Sensors 255

3.1 AuNPs-Based Amperometric Sensors 255

3.2 CNT-Based Amperometric Sensors 258

3.3 MNP-Based Amperometric Biosensors 262

3.4 Potentiometric Sensors 265

3.5 Conductometric and Impedimetric Biosensors 265

4 Conception and Modeling of Amplification Effect in Nanomaterial-Based Immunosensors 267

4.1 AuNP-Based Amperometric Immunosensors 267

4.2 AuNP-Based Potentiometric Sensors 272

4.3 Impedimetric Sensors 273

4.4 Conductometric Sensors 276

5 Conception and Modeling of Amplification Effect in Nanomaterial-Based DNA Biosensors 277

5.1 Amperometric Sensors 277
5.2 Impedimetric Sensors 283

6 Conclusion 284

References 285

8 Ion-Sensitive Field-Effect Transistors with Nanostructured Channels and Nanoparticle-Modified Gate Surfaces: Theory, Modeling, and Analysis **295**

V. K. Khanna

1 Introduction 295

2 Structural Configurations of the Nanoscale ISFET 297
2.1 The Nanoporous Silicon ISFET 297
2.2 The CNT ISFET 298
2.3 The Si-NW ISFET 299

3 Physics of the Si-NW Biosensor 299
3.1 Basic Principle 299
3.2 Analogy with the Nanocantilever 300
3.3 Preliminary Analysis of Micro-ISFET Downscaling to Nano-ISFET 301
3.4 Single-Gate and Dual-Gate Nanowire Sensors 304
3.5 Energy-Band Model of the NW Sensor 305

4 Nair-Alam Model of Si-NW Biosensors 307
4.1 The Three Regions in the Biosensor 307
4.2 Computational Approach 308
4.3 Effect of Nanowire Diameter (d) on Sensitivity at Different Doping Densities, with Air as the Surrounding Medium 310
4.4 Effect of Nanowire Length (L) on Sensitivity at Different Doping Densities, with Air as the Surrounding Medium 310
4.5 Effect of the Fluidic Environment 310
4.6 Overall Model Implications 315

5 pH Response of Silicon Nanowires in Terms of the Site-Binding and Gouy-Chapman-Stern Models 316

6 Subthreshold Regime as the Optimal Sensitivity Regime of Nanowire Biosensors 321

7 Effective Capacitance Model for Apparent Surpassing of the Nernst Limit by Sensitivity of the Dual-Gate NW Sensor 324

8 Tunnel Field-Effect Transistor Concept 326

9 Role of Nanoparticles in ISFET Gate Functionalization 328

9.1 Supportive Role of Nanoparticles 328
9.2 Direct Reactant Role of Nanoparticles 330

10 Neuron-CNT (Carbon Nanotube) ISFET Junction Modeling 332

11 Conclusions and Perspectives 334

Dedication 335

Acknowledgments 335

References 335

9 Biosensors: Modeling and Simulation of Diffusion-Limited Processes 339

L. Rajendran

1 Introduction 339

1.1 Enzyme Kinetics 339
1.2 Basic Scheme of Biosensors 340
1.3 The Nonlinear Reaction-Diffusion Equation and Biosensors 340
1.4 Types of Biosensors 342
1.5 Michaelis-Menten Kinetics 343
1.6 Non–Michaelis-Menten Kinetics 343
1.7 Importance of Modeling and Simulation of Biosensors 344

2 Modeling of Biosensors 345

2.1 Michaelis-Menten Kinetics and Potentiometric Biosensors 345
2.2 Michaelis-Menten Kinetics and Amperometric Biosensors 346
2.3 Michaelis-Menten Kinetics and Amperometric Biosensors for Immobilizing Enzymes 348
2.4 Michaelis-Menten Kinetics and the Two-Substrate Model 349
2.5 Non–Michaelis-Menten Kinetics 353
2.6 Other Enzyme Reaction Mechanisms 356
2.7 Kinetics of Enzyme Action 361
2.8 Trienzyme Biosensor 362

3 Microdisk Biosensors 363

3.1 Introduction 363
3.2 Mathematical Formulation of the Problem 364
3.3 First-Order Catalytic Kinetics 366
3.4 Zero-Order Catalytic Kinetics 370
3.5 For All Values of K_M 372
3.6 Conclusions 373

4 Microcylinder Biosensors 373

4.1 Introduction 373
4.2 Mathematical Formulation of the Problem 374
4.3 Analytical Solutions of the Concentrations and Current 376

4.4 Comparison with Limiting Case of Rijiravanich's Work 378
4.5 Discussion 379
4.6 Conclusions 381
4.7 PPO-Modified Microcylinder Biosensors 382

5 Spherical Biosensors 383
5.1 Simple Michaelis-Menten and Product Competitive Inhibition Kinetics 383
5.2 Immobilized Enzyme for Spherical Biosensors 385
5.3 Conclusion 386

Appendix: Various Analytical Schemes for Solving Nonlinear Reaction Diffusion Equations 386
A. Basic Concept of the Variational Iteration Method 386
B. Basic Concept of the Homotopy Perturbation Method 387
C. Basic Concept of the Homotopy Analysis Method 388
D. Basic Concept of the Adomian Decomposition Method 391

References 392

INDEX **399**

PREFACE

This series, *Chemical Sensors: Simulation and Modeling,* is the perfect complement to Momentum Press's six-volume reference series, *Chemical Sensors: Fundamentals of Sensing Materials* and *Chemical Sensors: Comprehensive Sensor Technologies,* which present detailed information about materials, technologies, fabrication, and applications of various devices for chemical sensing. Chemical sensors are integral to the automation of myriad industrial processes and everyday monitoring of such activities as public safety, engine performance, medical therapeutics, and many more.

Despite the large number of chemical sensors already on the market, selection and design of a suitable sensor for a new application is a difficult task for the design engineer. Careful selection of the sensing material, sensor platform, technology of synthesis or deposition of sensitive materials, appropriate coatings and membranes, and the sampling system is very important, because those decisions can determine the specificity, sensitivity, response time, and stability of the final device. Selective functionalization of the sensor is also critical to achieving the required operating parameters. Therefore, in designing a chemical sensor, developers have to answer the enormous questions related to properties of sensing materials and their functioning in various environments. This five-volume comprehensive reference work analyzes approaches used for computer simulation and modeling in various fields of chemical sensing and discusses various phenomena important for chemical sensing, such as surface diffusion, adsorption, surface reactions, sintering, conductivity, mass transport, interphase interactions, etc. In these volumes it is shown that theoretical modeling and simulation of the processes, being a basic for chemical sensor operation, can provide considerable assistance in choosing both optimal materials and optimal configurations of sensing elements for use in chemical sensors. The theoretical simulation and modeling of sensing material behavior during interactions with gases and liquid surroundings can promote understanding of the nature of effects responsible for high effectiveness of chemical sensors operation as well. Nevertheless, we have to understand that only very a few aspects of chemistry can be computed exactly.

However, just as not all spectra are perfectly resolved, often a qualitative or approximate computation can give useful insight into the chemistry of studied phenomena. For example, the modeling of surface-molecule interactions, which can lead to changes in the basic properties of sensing materials, can show how these steps are linked with the macroscopic parameters describing the sensor response. Using quantum mechanics calculations, it is possible to determine parameters of the energetic (electronic) levels of the surface, both inherent ones and those introduced by adsorbed species, adsorption complexes, the precursor state, etc. Statistical thermodynamics and kinetics can allow one to link those calculated surface parameters with surface coverage of adsorbed species corresponding to real experimental conditions (dependent on temperature, pressure, etc.). Finally, phenomenological modeling can tie together theoretically calculated characteristics with real sensor parameters. This modeling may include modeling of hot platforms, modern approaches to the study of sensing effects, modeling of processes responsible for chemical sensing, phenomenological modeling of operating characteristics of chemical sensors, etc.. In addition, it is necessary to recognize that in many cases researchers are in urgent need of theory, since many experimental observations, particularly in such fields as optical and electron spectroscopy, can hardly be interpreted correctly without applying detailed theoretical calculations.

Each modeling and simulation volume in the present series reviews modeling principles and approaches particular to specific groups of materials and devices applied for chemical sensing. *Volume 1: Microstructural Characterization and Modeling of Metal Oxides* covers microstructural characterization using scanning electron microscopy (SEM), transmission electron spectroscopy (TEM), Raman spectroscopy, in-situ high-temperature SEM, and multiscale atomistic simulation and modeling of metal oxides, including surface state, stability, and metal oxide interactions with gas molecules, water, and metals. *Volume 2: Conductometric-Type Sensors* covers phenomenological modeling and computational design of conductometric chemical sensors based on nanostructured materials such as metal oxides, carbon nanotubes, and graphenes. This volume includes an overview of the approaches used to quantitatively evaluate characteristics of sensitive structures in which electric charge transport depends on the interaction between the surfaces of the structures and chemical compounds in the surroundings. *Volume 3: Solid-State Devices* covers phenomenological and molecular modeling of processes which control sensing characteristics and parameters of various solid-state chemical sensors, including surface acoustic wave, metal-insulator-semiconductor (MIS), microcantilever, thermoelectric-based devices, and sensor arrays intended for "electronic nose" design. Modeling of nanomaterials and nanosystems that show promise for solid-state chemical sensor design is analyzed as well. *Volume 4: Optical Sensors* covers approaches used for modeling and simulation of various types of optical sensors such as fiber optic, surface plasmon resonance, Fabry-Pérot interferometers, transmittance in the mid-infrared region, luminescence-based devices, etc. Approaches used for design and optimization of optical systems aimed for both remote gas sensing and gas analysis chambers for the nondispersive infrared (NDIR) spectral range are discussed as well. A description of multiscale atomistic simulation of hierarchical nanostructured materials for optical chemical sensing is also included in this volume. *Volume 5: Electrochemical Sensors* covers modeling and simulation of electrochemical processes in both solid and liquid electrolytes, including charge separation and transport (gas diffusion, ion diffusion) in membranes, proton–electron transfers, electrode reactions, etc. Various models used to describe electrochemical sensors such as potentiometric, amperometric, conductometric, impedimetric, and ion-sensitive FET sensors are discussed as well.

I believe that this series will be of interest of all who work or plan to work in the field of chemical sensor design. The chapters in this series have been prepared by well-known persons with high qualification in their fields and therefore should be a significant and insightful source of valuable information for engineers and researchers who are either entering these fields for the first time, or who are already conducting research in these areas but wish to extend their knowledge in the field of chemical sensors and computational chemistry. This series will also be interesting for university students, post-docs, and professors in material science, analytical chemistry, computational chemistry, physics of semiconductor devices, chemical engineering, etc. I believe that all of them will find useful information in these volumes.

G. Korotcenkov

About the Editor

Ghenadii Korotcenkov received his Ph.D. in Physics and Technology of Semiconductor Materials and Devices in 1976, and his Habilitate Degree (Dr. Sci.) in Physics and Mathematics of Semiconductors and Dielectrics in 1990. For a long time he was a leader of the scientific Gas Sensor Group and manager of various national and international scientific and engineering projects carried out in the Laboratory of Micro- and Optoelectronics, Technical University of Moldova. Currently, Dr. Korotcenkov is a research professor at the Gwangju Institute of Science and Technology, Republic of Korea.

Specialists from the former Soviet Union know Dr. Korotcenkov's research results in the field of study of Schottky barriers, MOS structures, native oxides, and photoreceivers based on Group III–V compounds very well. His current research interests include materials science and surface science, focused on nanostructured metal oxides and solid-state gas sensor design. Dr. Korotcenkov is the author or editor of 11 books and special issues, 11 invited review papers, 17 book chapters, and more than 190 peer-reviewed articles. He holds 18 patents, and he has presented more than 200 reports at national and international conferences.

Dr. Korotcenkov's research activities have been honored by an Award of the Supreme Council of Science and Advanced Technology of the Republic of Moldova (2004), The Prize of the Presidents of the Ukrainian, Belarus, and Moldovan Academies of Sciences (2003), Senior Research Excellence Awards from the Technical University of Moldova (2001, 2003, 2005), a fellowship from the International Research Exchange Board (1998), and the National Youth Prize of the Republic of Moldova (1980), among others.

CONTRIBUTORS

Nikolai F. Uvarov (Chapter 1)
Institute of Solid State Chemistry and Mechanochemistry
Siberian Branch of the Russian Academy of Sciences
Novosibirsk 630128, Russia

Chongook Park (Chapter 2)
Department of Materials Science & Engineering
KAIST
Dae-jeon 305-701, South Korea

Inkun Lee (Chapter 2)
Department of Materials Science & Engineering
KAIST
Dae-jeon 305-701, South Korea

Dearo Lee (Chapter 2)
Department of Materials Science & Engineering
KAIST
Dae-jeon 305-701, South Korea

Jeffrey Fergus (Chapter 2)
Materials Research and Education Center
Auburn University
Auburn, Alabama 36849-5341, USA

Norio Miura (Chapter 2)
Art, Science and Technology Center for Cooperative Research
Kyushu University
Fukuoka 816-8580, Japan

Hyungjun Yoo (Chapter 2)
Department of Electrical Engineering
KAIST
Dae-jeon 305-701, South Korea

Antonio Ángel Moya Molina (Chapter 3)
Departamento de Física
Universidad de Jaén, Campus de las Lagunillas
Jaén 23071, Spain

Raluca-Ioana Stefan-van Staden (Chapter 4)
Laboratory of Electrochemistry and PATLAB Bucharest
National Institute of Research for Electrochemistry and Condensed Matter
Bucharest 060021, Romania

Konstantin N. Mikhelson (Chapter 5)
Ionometry Laboratory, Chemical Faculty
St. Petersburg State University
St. Petersburg, Russia

Sergio Bermejo (Chapter 6)
Department of Electronic Engineering
Universitat Politècnica de Catalunya (UPC)
Barcelona 08034, Spain

Nicole Jaffrezic-Renault (Chapter 7)
Institute of Analytical Chemistry, UMR CNRS 5280
Claude Bernard University Lyon 1
Villeurbanne 69100, France

Vinod Kumar Khanna (Chapter 8)
MEMS & Microsensors
CSIR—Central Electronics Engineering Research Institute
Pilani 333031 (Rajasthan), India

L. Rajendran (Chapter 9)
Department of Mathematics
The Madura College (Autonomous)
Madurai 625011, Tamil Nadu, South India

PART 3

Electrochemical Biosensors

CHAPTER 7

Nanomaterial-Based Electrochemical Biosensors

N. Jaffrezic-Renault

1. INTRODUCTION

Nowadays nanotechnology is sharing knowledge, tools, techniques, and information with electrochemistry and electroanalysis in other fields. Nanobiomaterials are one of the very important products of nanotechnology. Nanobiomaterials can be obtained, in general, by either the controlled assembly of nanoscale building blocks (a bottom-up approach) or controlled elimination of starting materials and biomaterials to the nanoscale (a top-down approach).

Certain nanomaterials are attractive probe candidates because of their (1) small size (1–100 nm) and correspondingly large surface-to-volume ratio, (2) chemically tailorable physical properties, directly related to size, composition, and shape, (3) unusual target binding properties, and (4) overall structural robustness.

Nanomaterials such as nanoparticles or carbon nanotubes connected with biomolecules, in the same size order of magnitude, are being used for several bioanalytical applications. Electroanalysis is taking advantage of all the possibilities offered by nanomaterials which are easy to detect by conventional electrochemical methods (i.e., electroactive nanoparticles, etc.) or compatible with (bio)sensor building technologies.

The most important advantages that nanomaterials bring to electroanalysis are the following.

DOI: 10.5643/9781606505984/ch7

1. Their immobilization on electrode surfaces generates a roughened conductive-high-surface-area interface that enables the sensitive electrochemical detection of molecular and biomolecular analytes.
2. They can act as effective labels for the amplified electrochemical analysis of the respective analytes.
3. The conductivity properties of metal nanoparticles enable the design of biomaterial architectures with predesigned and controlled electrochemical functions.

In this chapter, the state-of-the-art of nanomaterial-based electrochemical sensors will be presented, based on different bottom-up approach building technologies of nanomaterials, applied to different transducing techniques (amperometry, potentiometry, impedancemetry, and conductimetry), and to different biomolecules (enzymes, antibodies, DNA). Nanomaterials are classified as conducting nanomaterials (metallic nanoparticles such as gold nanoparticles and carbon nanotubes) and nonconducting nanomaterials such as magnetic nanoparticles (MNPs).

2. NANOMATERIALS: FABRICATION, CHEMICAL AND PHYSICAL PROPERTIES

2.1. CONDUCTING NANOMATERIALS

2.1.1. Metal Nanoparticles

Gold nanoparticles (AuNPs) can be prepared using an electron beam or through optical exposure. However, two more simple modes are generally preferred for the construction of electrochemical biosensors. In the first one, colloidal gold particles are prepared with a wide range of diameters and relatively high monodispersity by adding sodium citrate solution to a boiling $HAuCl_4$ aqueous solution (Katz et al. 2004). The size of the resulting colloidal gold nanoparticles, whose surfaces are negatively charged with citrate, is controlled by the molar ratio of $HAuCl_4$/sodium citrate (the lower the ratio, the smaller the particle size). The second method is based on the electrodeposition of gold nanoparticles from a $HAuCl_4$ solution onto the bulk electrode material (Rashid et al. 2006), and in this case experimental variables involved in the electrodeposition process (such as the applied potential and the time of deposition) govern the size and morphology of the nanoparticles formed.

Electrical properties of AuNPs have been characterized by physical measurements. Gold clusters stabilized by chemisorbed monolayers of octane-, dodecane- or hexadecanethiolate have been studied in the solid phase (Terrill et al. 1995). Films of the dry, solid cluster compound on interdigitated array electrodes exhibit current–potential responses, characteristic of electron hopping conductivity in which electrons tunnel from Au core to Au core. The electron hopping rate decreases and the activation barrier increases systematically at longer alkane chain length. The results are consistent with electron-transport rate control being a combination of thermally activated electron transfer to create oppositely charged Au cores (cermet theory) and distance-dependent tunneling through the oriented alkanethiolate layers separating them.

The ability of gold nanoparticles to provide a stable immobilization of biomolecules while retaining their bioactivity is a major advantage for the preparation of biosensors. Furthermore, gold nanoparticles permit direct electron transfer between redox proteins and bulk electrode materials, thus allowing electrochemical sensing to be performed with no need for electron transfer mediators. Characteristics of gold nanoparticles such as high surface-to-volume ratio, high surface energy, ability to decrease protein–metal particles distance, and the functioning as electron-conducting pathways between prosthetic groups and the electrode surface, have been claimed as reasons to facilitate electron transfer between redox proteins and electrode surfaces (Liu et al. 2003a; Yanez-Sedeno and Pingarron 2005). Gold nanoparticles have also been demonstrated to constitute useful interfaces for the electrocatalysis of redox processes of molecules such as H_2O_2, O_2, or NADH involved in many significant biochemical reactions (Rashid et al. 2006).

2.1.2. Carbon Nanotubes

Carbon nanotubes (CNTs) have become the subject of intense investigation since their discovery (Rao et al. 2001). CNTs can be made by chemical vapor deposition, carbon arc methods, or laser evaporation and can be divided into single-wall carbon nanotubes (SWCNTs) and multiwall carbon nanotubes (MWCNTs). SWCNTs possess a cylindrical nanostructure (with a high aspect ratio), formed by rolling up a single graphite sheet into a tube. MWCNTs comprise several layers of graphene cylinders that are concentrically nested like rings of a tree trunk (with an interlayer spacing of 0.34 nm (Baughman et al. 2002; Davis et al. 2003). The unique properties of carbon nanotubes make them extremely attractive for the task of chemical sensors in general, and electrochemical detection in particular (Zhao et al. 2002). Such considerable interest reflects the unique behavior of CNTs, including their remarkable electrical, chemical, mechanical, and structural properties. CNTs can display metallic, semiconducting, and superconducting electron transport, possess a hollow core suitable for storing guest molecules, and have the largest elastic modulus of any known material (Davis et al. 2003).

Recent studies have demonstrated that CNTs can enhance the electrochemical reactivity of important biomolecules (Zhao et al. 2002; Musameh et al. 2002), and can promote the electron transfer reactions of proteins (including those where the redox center is embedded deep within the glycoprotein shell) (Gooding et al. 2003; Yu et al. 2003). In addition to enhanced electrochemical reactivity, CNT-modified electrodes have been shown to be useful to accumulate important biomolecules (e.g., nucleic acids) (Wang et al. 2003a) and to alleviate surface fouling effects (such as those involved in the NADH oxidation process) (Musameh et al. 2002). The remarkable sensitivity of CNT conductivity to the surface adsorbates permits the use of CNTs as highly sensitive nanoscale sensors. These properties make CNTs extremely attractive for a wide range of electrochemical biosensors ranging from amperometric enzyme electrodes to DNA hybridization biosensors. To take advantage of the remarkable properties of these unique nanomaterials in such electrochemical sensing applications, the CNTs need to be properly functionalized and immobilized (Wang 2005).

2.2. NONCONDUCTING NANOMATERIALS: MAGNETIC NANOPARTICLES

Magnetic nanoparticles (MNPs) consist of a paramagnetic or superparamagnetic core surrounded by a polymeric outer layer suitable for the immobilization of biomolecules (Albers et al. 2003). The magnetic core is readily available in different iron oxide forms, among which magnetite (Fe_3O_4) and maghemite (γ-Fe_2O_3, ferrimagnetic) stand out because their compatibility has been proven in biolabeling and bioseparation (Andreescu et al. 2009; Teja and Koh 2009; Simon de Dios et al. 2010).

The development of coatings for the magnetic core of MNPs was necessary to address their limitations, including (1) high surface energies which lead to aggregation, (2) high chemical activity leading to their oxidation, loss of magnetic properties, and dispersibility when exposed to air, and (3) biodegradation with subsequent changes in magnetic properties. Consequently, protective shells, mainly composed of agarose, cellulose, silica, silicone, porous glass, mica, or polystyrene (Palecek and Fojta 2007), have been developed to protect and preserve the stability of iron oxide MNPs. These shells also allow further functionalization, thus promoting the performance of MNPs as recognition elements in sensing and (bio)chemical arrays (Simon de Dios et al. 2010). Moreover, nowadays there are commercially available MNPs modified with biomolecules that allow their use in different types of bioassays. So, MNPs functionalized with (1) streptavidin, suitable for capturing biotinylated nucleic acids, aptamers, peptides, proteins, etc., (2) protein A (protA) or protein G (protG), which specifically bind antibodies, (3) oligonucleotides, or (4) affinity ligands for specific capture of tagged recombinant proteins, etc. (Palecek and Fojta 2007), can be purchased from different companies. MNP modification has resulted in important practical advantages from an analytical point of view, including (1) shorter reaction times between dissolved species and biomolecules immobilized on the surface of the beads, which is also favored by the easy dispersion of MNPs into solution with only gentle shaking; (2) ready miniaturization of the assay system by using MNPs as a mobile solid phase; (3) reduction of the required volumes of reagents and produced waste; and (4) obtaining lower detection limits with shorter assay times (Kuramitz 2009).

The development and application of MNPs in separation and detection methodologies has attracted strong interest in the last years. This is mainly due to the versatility, high surface area, chemical and physical stability, low toxicity, and high biocompatibility exhibited by MNPs (Kuramitz 2009). Their size, similar to that of molecules in nature, range from nanometers to a few millimeters, and the particle-linked molecules can quickly agglomerate and be separated from a matrix or resuspended in an appropriate working medium without retaining any residual magnetism as a consequence of a change in an external magnetic force (Hsing et al. 2007; Simon de Dios et al. 2010).

Owing to all these properties, MNPs constitute an attractive platform for the design of electrochemical biosensors which also add their inherent advantages such as system miniaturization, low cost, and ease of operation (Hsing et al. 2007). In fact, electrochemical enzyme-based biosensors, immunosensors, and nucleic acid hybridization–based sensors have been described in the literature and cited in several reviews (Palecek and Fojta 2007; Aguilar-Arteaga et al. 2010; Dolatabadi et al. 2011).

3. CONCEPTION AND MODELING OF AMPLIFICATION EFFECT IN NANOMATERIAL-BASED ENZYME SENSORS

3.1. AuNPs-BASED AMPEROMETRIC SENSORS

Much of the research on biosensors involving gold nanoparticles has been devoted to enzyme electrodes.

3.1.1. Direct Enzyme Wiring with AuNPs (Willner et al. 2011)

The electrical contacting of redox enzymes with electrodes is a fundamental issue in bioelectrochemistry and is essential for the development of amperometric biosensors and biofuel cell elements (Heller 1990; Willner 2002). The direct electron transfer between the redox center embedded in the protein structure of the enzymes and the electrodes is, usually, prohibited due to the spatial separation of the redox center from the electrode. Electron transfer theory (Marcus and Suttin 1985) implies that the transfer rate K_{ct} between a donor and acceptor pair is dominated by the distance (d) separating the donor and acceptor units (where d_o is the van der Waals distance separating the components, ΔG° is the free-energy change accompanying the electron transfer, and λ is the reorganization energy involved in the electron transfer process). K_{ct} is proportional to the following expression (Eq. 7.1), where β is a proportionality factor:

$$K_{ct} \propto \exp[-\beta(d - d_o)] \times \exp[-(\Delta G^\circ + \lambda)^2 / (4RT\lambda)] \tag{7.1}$$

Thus, for effective electrical contacting of the redox center of the biocatalyst with the electrode, shortening of the electron transfer distances is important, and the alignment of the redox center as close as possible to the electrode surface is essential.

An AuNP (diameter 1.4 nm) functionalized with an active N-hydroxysuccinimide functionality was modified with an amino flavin adenine dinucleotide, amino-FAD, (1), cofactor unit, Figure 7.1. Apo-glucose oxidase (apo-GOx), which lacks its native FAD cofactor, was then reconstituted on the FAD-functionalized Au NPs (Xiao et al. 2003). The GOx-reconstituted Au NPs were then assembled on Au electrodes, using different dithiol linkers. The enzyme-modified electrodes revealed electrical contact between the redox center of the biocatalyst and the electrode, and bioelectrocatalytic oxidation of glucose was activated, thus providing amperometric sensing of the sugar. Knowing the surface coverage of the enzyme and the saturation current generated by the enzyme electrode, the turnover rate of electrons between the active site and the electrode corresponded to $5 \times 10^3\ s^{-1}$. This exchange rate of electrons between the enzyme and the electrode is ca. sevenfold higher than the rate of electron transfer from the active center to the native oxygen (O_2) acceptor (Bourdillon et al. 1993). The efficient electrical contacting of the redox center with the electrode leads to an oxygen-insensitive enzyme electrode,

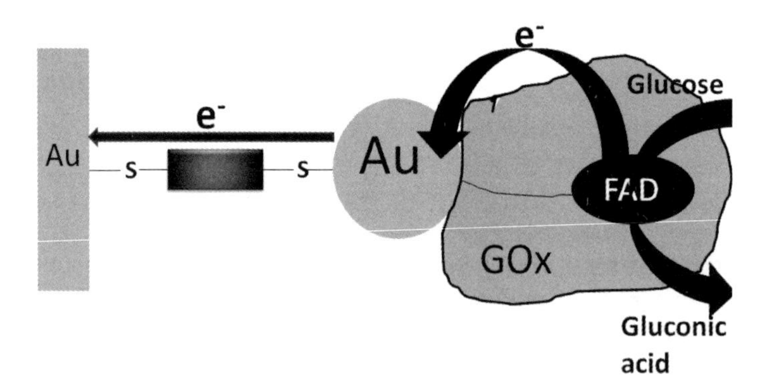

Figure 7.1. Assembly of Au-NP–reconstituted GOx electrode obtained or by the adsorption of Au-NP–reconstituted GOx to a dithiol monolayer associated with a Au electrode or by the adsorption of Au-NPs functionalized with FAD on the dithiol-modified Au electrode followed by the reconstitution of apo-GOx on the functional NPs.

and to a glucose sensor that is unperturbed by common electrochemically active interferants, such as ascorbic or uric acids.

3.1.2. Different Strategies for AuNPs and Enzyme Immobilization

3.1.2.1. DIRECT ELECTRODEPOSITION OF GOLD NANOPARTICLES

Among the various strategies followed, a useful and simple way consists of the direct electrodeposition of nanoparticles onto the electrode surface. For example, tyrosinase immobilization by cross-linking onto a glassy carbon electrode (GCE) modified with electrodeposited gold nanoparticles was used to prepare a biosensor which showed high activity toward various phenolic compounds (Carralero Sanz et al. 2005). Cyclic voltammograms obtained with and without AuNPs show that the presence of electrodeposited AuNPs increases the measured current induced by the enzymatic reaction by a factor of 3. Also, a glucose biosensor was prepared by covalent attachment of GOx to a gold nanoparticle monolayer–modified Au electrode (Zhang et al. 2005). Another example is the construction of a xanthine oxidase biosensor for the determination of hypoxanthine. This biosensor makes use of a carbon paste electrode (CPE) modified with electrodeposited gold nanoparticles, onto which the enzyme was cross-linked with glutaraldehyde and bovine serum albumin (BSA) (Agui et al. 2006). The XOD–nAu–CPE configuration allowed Hx detection to be carried out at 0.00 V, with the subsequent minimization of interferences such as ascorbic acid. The detection limit achieved for Hx, 2.2×10^{-7} M, was similar to the best reported values with biosensor designs without gold nanoparticles working at much more extreme potentials. Electrodeposition of gold nanoparticles onto a planar Au electrode was also used to create a favorable surface for the attachment of acetylcholinesterase. The enzyme-modified electrode was employed to detect nanomolar concentrations of the pesticide carbofuran (Shulga and Kirchhoff 2007).

3.1.2.2. MODIFICATION OF ELECTRODE SURFACES WITH SELF-ASSEMBLED MONOLAYERS

Modification of electrode surfaces with self-assembled monolayers (SAMs) of thiols provides a simple way to design tailored materials that can be further used as functionalized sites to immobilize gold nanoparticles and enzymes (Gooding and Hibbert 1999). A comparison of the analytical performance of different GOx biosensor designs based on several SAM-modified electrodes showed that a configuration involving colloidal gold bound to cysteamine monolayers self-assembled on a gold disk electrode exhibited high sensitivity and long biosensor lifetime in comparison with other GOx biosensors (Mena et al. 2005).

3.1.2.3. INCORPORATION OF AUNPS INTO COMPOSITE ELECTRODE MATRICES

Various research activities have been directed toward combining the advantageous features of colloidal gold and carbon paste electrodes. A reagentless glucose biosensor based on direct electron transfer of GOx (Liu and Ju 2003), a horseradish peroxidase (HRP) biosensor (Liu and Ju 2002), and a Tyr biosensor for phenol detection (Liu et al. 2003b) have been constructed by immobilizing the corresponding enzymes onto electrodes prepared by mixing colloidal gold with the carbon paste components. Based on a similar methodology, a composite electrode was also recently prepared by modifying glassy carbon microparticles with gold nanoparticles and xanthine oxidase for xanthine and hypoxanthine detection (Cubukcu et al. 2007).

Enzyme biosensors using composite graphite–Teflon electrodes modified with gold nanoparticles have been prepared by Pingarron's group. The biosensor design is based on a graphite–Teflon composite matrix in which the enzyme(s) and colloidal gold nanoparticles are physically incorporated. Based on this methodology, a tyrosinase biosensor with improved stability and sensitivity with respect to other configurations was fabricated. The Tyr–Aucoll–graphite–Teflon biosensor exhibited suitable amperometric responses at −0.10 V for different alkyl- and chlorophenols. The detection limits obtained were 3 nM for catechol and approximately 20 nM for other phenolic compounds. The presence of colloidal gold in the composite matrix gave rise to enhanced kinetics of both the enzyme reaction and the electrochemical reduction of the corresponding *o*-quinones at the electrode, thus providing high sensitivity. The biosensor exhibited excellent renewability by simple polishing, with a lifetime of at least 39 days without apparent loss of enzyme activity (Carralero et al. 2006).

3.2. CNT-BASED AMPEROMETRIC SENSORS

3.2.1. Direct Enzyme Wiring with CNTs

The possibility of direct electron transfer between CNTs and GOx paves the way for the construction of amperometric glucose biosensors. Such electrical communication with GOx (and other oxidoreductase enzymes) would obviate the need for a co-substrate and allow efficient transduction of the biorecognition event.

The redox centers of glucose oxidase, like those of most oxidoreductases, are electrically insulated by a protein shell. Because of this glycoprotein shell, the enzyme cannot be oxidized or reduced at an electrode at any potential. Guiseppi-Elie et al. (2002) reported on direct electron transfer between SWCNTs and the redox center of adsorbed GOx. Both FAD and GOx were found to spontaneously

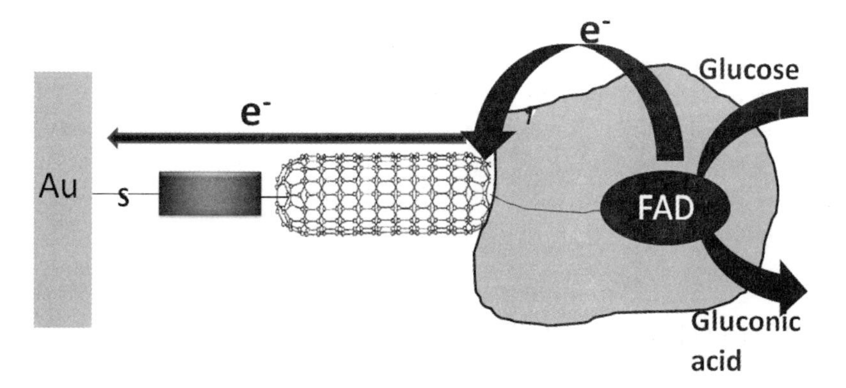

Figure 7.2. Assembly of the SWCNT electrically contacted with glucose oxidase on a gold electrode surface.

adsorb to annealed CNTs that were cast onto the glassy carbon surface to display quasi-reversible one-electron transfer. Similarly, GOx was found to spontaneously adsorb to annealed SWCNTs and to display a quasi-reversible one-electron transfer. It was assumed that the tubular fibrils become positioned within a tunneling distance of the cofactors with little consequence to denaturation.

Willner's group (Patolsky et al. 2004) demonstrated that aligned reconstituted GOx on the edge of SWCNTs can be linked to an electrode surface (Figure 7.2).The electrons were transported along distances greater than 150 nm, with the length of the SWCNT controlling the rate of electron transport. Interfacial electron-transfer rate constants of 42 and 19 s^{-1} were estimated for 50- and 100-nm-long SWCNTs, respectively. Such enzyme reconstitution on the ends of CNTs represents an extremely efficient approach for plugging an electrode into GOx. Luong et al. (2004) reported on the promoted electron transfer of GOx at a MWCNT-modified glassy carbon electrode.

3.2.2. *CNT-Based Enzyme Biosensors*

The successful realization of CNT-based biosensors requires proper control of their chemical and physical properties, as well as their functionalization and surface immobilization. The purification of CNTs with oxidizing acids creates surface acid sites that are mainly carboxylic moieties. These functional groups provide sites for covalent linking of CNTs to biorecognition elements (or other materials) or for their integration onto polymer surface structures. Other derivatization methods include defect and noncovalent functionalization (Hirsch 2002). Enzymes can be immobilized on CNTs with enhanced retention of their biocatalytic activity. From the perspective of electrochemical reactivity, the side walls of CNTs are suggested to have properties similar to those of the basal plane of highly oriented pyrolytic graphite (HOPG), while their open ends resemble the edge planes of HOPG.

There are different ways for confining CNTs onto electrochemical transducers. Most commonly this is accomplished using CNT-coated electrodes (Musameh et al. 2002; Wang et al. 2003; Luong et al. 2004) or using CNT/binder composite electrodes (Wang and Musameh 2003a; Rubianes and Rivas 2003). Details of such CNT-based transducers are given hereafter.

3.2.2.1. CNT-COATED ELECTRODE TRANSDUCERS

One barrier for developing CNT film–based biosensing devices is the insolubility of CNTs in most solvents. Earlier reported CNT-modified electrodes thus commonly relied on casting a CNT/sulfuric acid solution onto a glassy carbon surface (Musameh et al. 2002). Nafion films have been used extensively for the construction of amperometric biosensors owing to their unique ion-exchange, discriminative, and biocompatibility properties. The ability of Nafion to solubilize CNT provides a useful avenue for preparing CNT-based electrode transducers for a wide range of sensing applications (Wang et al. 2003b). The resulting biosensors benefit greatly from the coupling of the antifouling/discriminative properties of Nafion films with the efficient electrocatalytic action of CNTs toward hydrogen peroxide. Further improvements in the electroactivity of hydrogen peroxide (with detection limit down to 25 nM) can be obtained by dispersing platinum nanoparticles in the Nafion/CNT coating (Hrapovic et al. 2004).

3.2.2.2. VERTICALLY ALIGNED NANOTUBES ON ELECTRODE ARRAYS

An effective CNT coating can be obtained by aligning short SWCNTs normal to an electrode by self-assembly. Such vertically aligned SWCNTs act as molecular wires to allow electrical communication between the underlying electrode and a redox enzyme (Gooding et al. 2003; Yu et al. 2003). Such direct electron transfer between the prosthetic group of the enzyme and an electron surface obviates the need for redox mediators and is thus extremely attractive for developing reagentless sensing devices. Specific biosensing applications of these CNT “forests” are discussed here. Similar arrays of aligned MWCNTs, grown on platinum substrates, were shown to be useful for amperometric enzyme electrodes (Sotiropoulou and Chaniotakis 2003). The carboxylated open ends of the nanotubes were used for the immobilization of the enzymes, while the platinum substrate provided the signal transduction. Lin and co-workers (Lin et al. 2004) developed amperometric biosensors based on CNT–nanoelectrode ensembles (NEEs). Such NEEs consist of millions of nanoelectrodes (each of less than 100 nm in diameter), embedded in a polymer-based epoxy, on a chromium-coated silicon substrate. The enzyme was immobilized on the CNT NEE using carbodiimide chemistry by forming amide linkages between its amine residues and carboxylic acid groups on the CNT tips. Koehne et al. (2003) fabricated low-density CNT arrays on silicon chips using a bottom-up approach, involving lithographic patterning, metallization of the electrical contacts, deposition of a catalyst, and CNT growth by plasma-enhanced chemical vapor deposition.

3.2.2.3. CNT-BASED BIOCOMPOSITE ELECTRODES

An attractive avenue for preparing CNT-based amperometric enzyme electrodes involves CNT/insulator/enzyme biocomposites (Wang and Musameh 2003b; Rubianes and Rivas 2003; Valentini et al. 2003). Conventional carbon paste composites have been widely used for the design of renewable amperometric enzyme electrodes (Gorton 1995). Britto et al. (1996) prepared a composite electrode based on mixing MWCNTs and bromoform as a binder. Carbon nanotube paste enzyme electrodes were prepared by mixing CNT with mineral oil (Rubianes and Rivas 2003; Valentini et al. 2003). Such composite electrodes combine the ability of carbon nanotubes to promote electron transfer reactions with the attractive advantages of paste electrode materials. The preparation of a binderless biocomposite based on mixing an enzyme (GOx) within the CNTs was reported (Wang and Musameh 2003a). The resulting biocomposite was packed within a needle and was used as a microsensor for glucose. Another simple avenue for preparing effective CNT-based electrochemical biosensors involves the use of CNT–Teflon composite materials (Wang and Musameh 2003b). These biocomposite devices rely on the use of CNTs as the sole conductive component of the transducer rather than utilizing them as the modifier in connections to another electrode surface. The bulk of the resulting CNT–Teflon electrodes serve as a "reservoir" of the enzyme. The strong electrocatalytic activity of CNTs toward hydrogen peroxide or NADH is not impaired by their association with the Teflon binder.

Screen-printed strip electrodes, based on thick-film microfabrication, offer large-scale mass production of highly reproducible, low-cost electrochemical biosensors. It has been demonstrated that CNT-based inks are highly suitable for the microfabrication of thick-film electrochemical sensors (Wang and Musameh 2004). Such screen-printed CNT sensors have a well-defined appearance, are mechanically stable (with good resistance to mechanical abrasion), and exhibit higher electrochemical reactivity (compared to conventional carbon strips). The resulting devices thus combine the attractive advantages of CNT materials and screen-printed electrodes, offer improved performance compared to strips fabricated with conventional carbon inks, and open the door for a wide range of sensing applications.

3.2.2.4. HYBRID MATERIALS WITH CARBON NANOTUBES

Conjugation of gold nanoparticles with other nanomaterials and biomolecules is an attractive research area within nanobiotechnology (Besteman et al. 2003). Hybrid nanoparticles/nanotubes materials have been shown to possess interesting properties, which can be applied for the development of electrochemical biosensors. A colloidal gold–CNT composite electrode using Teflon as the nonconducting binding material showed significantly improved responses to H_2O_2 when compared with other carbon composite electrodes, including those based on CNTs.

The incorporation of GOx into the new composite matrix allowed the preparation of a mediatorless glucose biosensor with a remarkably higher sensitivity than that from other GOx–CNT bioelectrodes (Ye et al. 2004).

3.3. MNP-BASED AMPEROMETRIC BIOSENSORS

In parallel with biosensing applications, enzyme-coated magnetic nanoparticles have been used to facilitate the enzyme handling. In this way, stock solutions containing such a material can be, when properly stored, used for more than 10 months without significant activity loss. This immobilization method has been validated for glucose oxidase, urease, and alpha-amylase covalently immobilized onto polyacrolein beads (Varlan et al. 1996).

Except for the work of Miyabayashi et al. (1988), where biomodified MNPs were used on a Clark electrode for glucose or *Saccharomyces cerevisiae* detection, one of the first applications of MNPs for biosensors was the combination of a covalent enzyme bonding onto MNPs with physical entrapment on the sensor surface. Latex nanoparticles, containing superparamagnetic material covalently modified with enzymes, have been patterned on a transducer by the means of screen-printed thick-films permanent magnet. Such a method, valuable for batch production independent of the nature of the substrate material, was evaluated for assessing glucose concentrations up to 20 mM (Varlan et al. 1995).

Enzymes immobilized on magnetic microparticles can be trapped by magnets and retained on an electrode surface at a specific location in flow analysis devices. However, only a few works using magnetic microparticles have been devoted to designing real enzymatic biosensors. As an example of such devices, an enzyme-based electrochemical magnetobiosensor for environmental toxicity analysis can be cited. The integration of the bioactive material as urease, cholinesterases, significantly increases the sensitivity and allows detection limits as low as 10^{-11} M for heavy-metal ions and 10^{-12} M for organophosphates and carbamates (Sole et al. 2001). In Table 7.1, characteristics of some magnetic microbeads–based enzymatic biosensors are gathered.

The potentialities of nanoporous silica microparticles containing superparamagnetic defect spinel-type iron oxide nanoparticles inside their pores have been

Table 7.1. Examples of magnetic nanoparticle-based enzymatic biosensors

Analyte	Magnetic nanoparticles	Transducer	Enzyme	Sensitivity (mA/mM)	Dynamic range (mM)	Ref.
Phenol	$MgFe_2O_4$ silica coated $\phi = 120$ nm	C paste	Tyrosinase	54.2	10^{-3}–0.5	Liu et al. 2005
Ethanol	Precipitated Fe_3O_4 $\phi = 9.8$ nm	Screen-printed C paste	Yeast YADH/NAD^+	0.61	1–9	Liao et al. 2006
Glucose	Precipitated Fe_3O_4 $\phi = 9.8$ nm	Screen-printed C paste	Glucose oxidase	1.74×10^3	0–33	Lu et al. 2006

underlined for biosensors applications. This approach allows obtaining high sensitivity and selectivity even in complex media. The interest of such a support has been validated for the immobilization of horseradish peroxidase leading to hydroquinone and hydrogen peroxide biosensing at levels as low as 4×10^{-7} M (Elyacoubi et al. 2006). The design of a magnetized carbon paste electrode with trapped magnetic nanoporous silica microparticles (MMPs) is given in Figure 7.3.

A similar strategy has been used for the immobilization of horseradish peroxidase with a high density of nanopores of MMPs having a diameter of 5 μm and applied for studying peroxidation of clozapine. Clozapine is a drug belonging to the dibenzoazepine class and is often used to treat neurological disorders. The resulting amperometric biosensor allows drug quantification in the micromolar range and presents quite good stability, no significant signal loss being observed after one month of storage (Yu et al. 2006).

Enzymatic MMP-based devices also present specific advantages, as MMPs do not act as a barrier for the diffusion of the analyte to the electrode surface. A decrease of only few percent of the voltamperometric signal is observed in the presence of microparticles compared to the bare electrode. Furthermore, MMPs constitute a valuable tool for inhibition studies, immobilized enzymes on such macroparticles being less sensitive, through a screening effect, to inhibitors than soluble enzymes. Thus, it has been recently shown that immobilized HRP, using an MMP strategy, was protected from inactivation by inhibitors such as thiols,

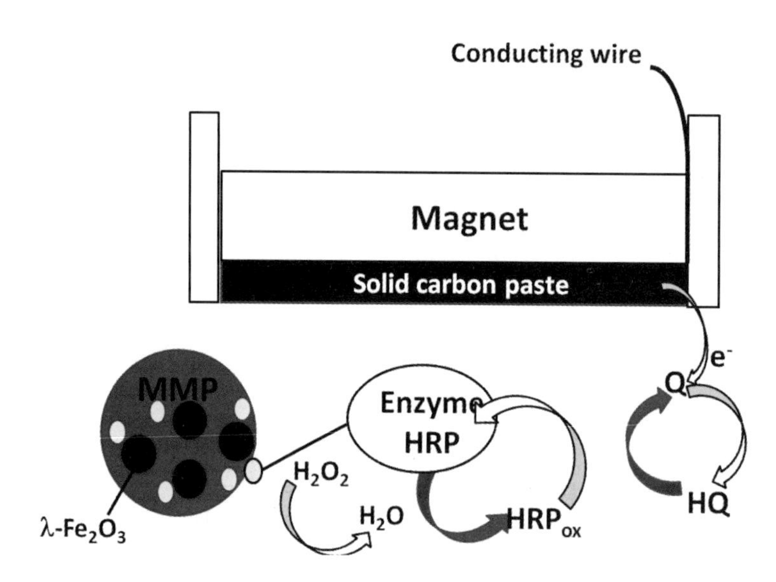

Figure 7.3. An enzymatic biosensor based on magnetic nanoporous silica microparticles, where oxidation of hydroquinone (HQ) to quinone by the enzyme horseradish peroxidase (HRP) in presence of hydrogen peroxide and subsequent electroreduction of quinone (Q) is shown.

which can react with intermediary quinoneimine derivatives produced during the enzymatic reaction (Kauffmann et al. 2006).

3.4. POTENTIOMETRIC SENSORS

Functionalization of field-effect (bio)chemical sensors with layer-by-layer (LbL) films containing single-walled carbon nanotubes and polyamidoamine (PAMAM) dendrimers has been reported (Siquiera et al. 2009a). A capacitive electrolyte–insulator–semiconductor (EIS) structure modified with carbon nanotubes (EIS-NTs) was built, which could be used as a penicillin biosensor. From atomic force microscopy (AFM) and field-emission scanning electron microscopy (FESEM) images, the LbL films were shown to be highly porous due to interpenetration of SWNTs into the dendrimer layers. Capacitance–voltage (C/V) measurements pointed to a high pH sensitivity of ca. 55 mV/pH for the EIS-NT structures. The biosensing ability toward penicillin of an EIS-NT-penicillinase biosensor was also observed, as the flat-band voltage shifted to lower potentials at different penicillin concentrations. A dynamic response of penicillin concentrations, ranging from 5.0 μM to 25 mM, was evaluated for an EIS-NT with the penicillinase enzyme immobilized onto the surfaces, via constant–capacitance (ConCap) measurements, achieving a sensitivity of ca. 116 mV/decade. The presence of the nanostructured PAMAM/SWNT LbL film led to sensors with higher sensitivity and better performance.

Constant-current (CC) measurements showed that the incorporation of the PAMAM/SWNT LbL film containing up to six bilayers onto the LAPS structure has a high pH sensitivity of ca. 58 mV/pH. The biosensing ability of the devices was tested for penicillin G via adsorptive immobilization of the enzyme penicillinase atop the LbL film. LAPS architectures modified with the LbL film exhibited higher sensitivity, ca. 100 mV/decade, in comparison to ca. 79 mV/decade for an unmodified LAPS, which demonstrates the potential application of the CNT-LbL structure in field-effect–based (bio)chemical sensors (Siquiera et al. 2009b)

3.5. CONDUCTOMETRIC AND IMPEDIMETRIC BIOSENSORS

3.5.1. Biocatalytic Growth of Metallic Gold Nanoparticles

Recent studies have shown that enzymes can mediate growth of metallic nanoparticles (Hwang et al. 2005), and that thiocholine (Pavlov et al. 2005; Liao et al 2012), H_2O_2 (Zayats et al. 2005; Lim et al. 2010), nicotinamide adenine dinucleotide (phosphate) (Xiao et al. 2004), glucose, together with glucose oxidase (Xiao et al. 2005), some small active molecules as reductants have led to the enlargement of AuNPs in the presence or absence of gold nano-seeds.

The biocatalytic growth of metallic gold nanoparticles is possible by reducing gold tetrachloride, in contact of AuNPs seeds, by hydrogen peroxide [Eq. (7.2)] produced by the enzyme glucose oxidase, in presence of its substrate [Eq. (7.3)].

$$AuCl_4^- + \frac{3}{2}H_2O_2 \xrightarrow{\text{AuNP seeds}} Au^\circ \downarrow + 4Cl^- + 3H^+ + \frac{3}{2}O_2 \quad (7.2)$$

$$\text{Glucose} + O_2 + H_2O \xrightarrow{\text{glucose oxidase}} H_2O_2 + \text{gluconolactone} \quad (7.3)$$

The size and density of gold nanoparticles are increased when concentration of glucose increases. This concept has been demonstrated and used for biosensing in homogeneous phase and in immobilized phase using absorbance spectroscopy by Baron and Willner (Zayats et al. 2005; Baron et al. 2005). It has been also implemented for sensitive voltammetric detection of glucose (Yan et al. 2008; Zhang et al. 2011) and for its impedimetric detection (Bourigua et al. 2012). Obtained detection limits of glucose are in the range of a few micromolar for all the transducing techniques involving biocatalytic growth of gold nanoparticles.

3.5.2. AuNPs Functionalized with Enzymes Using LbL Technique

A new conductometric biosensor based on interdigitated electrodes (IDEs) has been developed for the detection of urea using gold nanoparticles, synthesized using the citrate process, with an average diameter of 23 nm and functionalized with urease using layer-by-layer technique. Due to the ions generated by the hydrolysis of urea catalyzed by urease according to Eq. (7.4),

$$CO(NH_2)_2 + H^+ + 2H_2O \xrightarrow{\text{Urease}} 2NH_4^+ + HCO_3^- \quad (7.4)$$

a significant increase of local conductivity can be observed, as demonstrated by Jdanova et al. (1996).

The interdigitated electrodes allow the measurement of the change of conductivity in the region defined by field lines. The thickness involved is of the order of the interdigit distance (a few tens of micrometers) (Pänke et al. 2008). As has been modeled (Sheppard et al. 1996; Temple-Boyer et al. 2008), the observed steady-state response of the conductometric biosensor is the result of the reaction rate–limited kinetics of the enzymatic reaction and the diffusive flux of urea hydrolysis products away from the transducer surface, in the boundary layer. The most representative result of enzymatic reaction is pH variation; a pH limit value is reached at the transducer surface, and pH increase is very deep close to the electrode surface (Temple-Boyer et al. 2008).

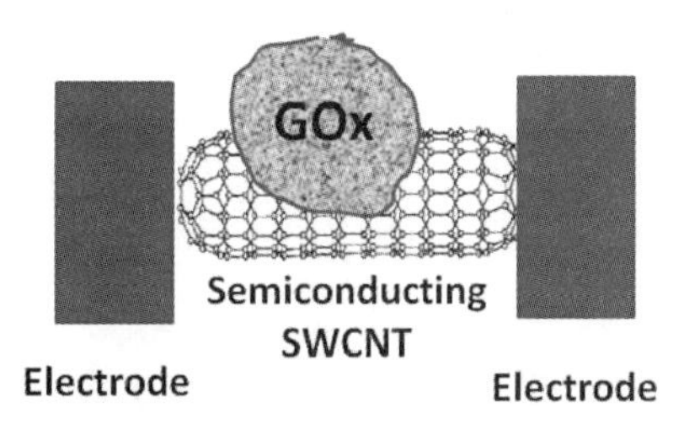

Figure 7.4. Schematic picture of two electrodes connecting a semiconducting SWCNT with GOx enzymes immobilized on its surface.

A detection limit of 100 μM of urea is obtained when cross-linked urease is directly immobilized on top of the IDEs (interdigitated distance: 20 μm), whereas a detection limit of 2 μM is obtained when urease-functionalized gold nanoparticles are deposited on top of the IDEs. The use of gold nanoparticles allows an increase of the sensitivity of detection (from 10 μS/mM to 107 μS/mM) due to the decrease of the thickness of probed zone (Nouira et al. 2012).

A glucose conductometric biosensor was developed using two types of nanoparticles (gold and magnetic), glucose oxidase (GOD) being adsorbed on poly(allylamine hydrochloride) (PAH)–modified nanoparticles, deposited on a planar interdigitated electrode. The best sensitivities for glucose detection were obtained with magnetic nanoparticles (70 μS/mM and 3 μM of detection limit) compared to 45 μS/mM and 9 μM with gold nanoparticles and 30 μS/mM and 50 μM with GOD directly cross-linked on IDEs (Nouira et al. 2013). This work shows that the amplification effect depends on the type of enzymatic reaction.

3.5.3. A SWCNT Transistor-Based Biosensor

Besteman et al. (2003) reported on a SWCNT transistor as a conductivity glucose biosensor (Figure 7.4). Controlled attachment of GOx to the SWCNT side wall was achieved through a linking molecule. Such immobilization led to decreased conductivity. The enzyme-coated tube acted as a reversible pH sensor and displayed an increase in conductance upon adding glucose (while applying a gate voltage between a standard reference electrode and the semiconducting SWCNT).

4. CONCEPTION AND MODELING OF AMPLIFICATION EFFECT IN NANOMATERIAL-BASED IMMUNOSENSORS

4.1. AuNP-BASED AMPEROMETRIC IMMUNOSENSORS

This section summarizes recent approaches in the construction of electrochemical immunosensors in which gold nanoparticles play a crucial role both in the enhancement of the electrochemical signal transducing the binding reaction of antigens at antibody-immobilized surfaces and in the ability to increase the amount of immobilized immunoreagents in a stable mode.

4.1.1. Labeling with HRP

A different strategy was used to fabricate an amperometric immunosensor based on immobilization of hepatitis B antibody on a gold electrode modified with gold nanoparticles and horseradish peroxidase (Zhuo et al. 2005b). In this immunosensor, HRP was incorporated instead of BSA to block the possible remaining active sites of the gold nanoparticles monolayer, in order to avoid nonspecific adsorption and also amplify the response of the antigen–antibody reaction. Figure 7.5 shows the steps involved in the immunosensor preparation.

α-Fetoprotein (AFP), an oncofetal glycoprotein, is a widely used tumor marker for the diagnosis of patients with germ-cell tumors and hepatocellular carcinoma. An immunosensor for AFP was prepared by entrapping thionine (Thi) into Nafion (Nf) to form a composite Thi/Nf membrane, which yields an interface containing amine groups to assemble gold nanoparticle layers for immobilization of α-1-fetoprotein antibody. After the immunosensor was incubated with AFP, the CVs current decreased linearly in concentration ranges of AFP (Zhuo et al.

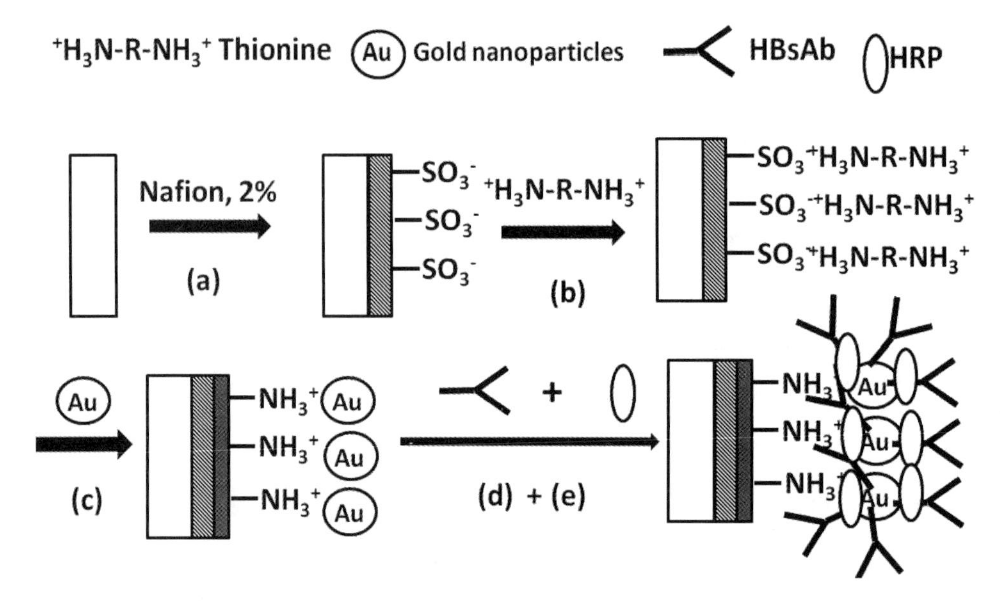

Figure 7.5. Scheme showing the stepwise HBsAg immunosensor preparation: (a) formation of a Nafion monolayer; (b) adsorption of thionine; (c) formation of gold nanoparticles monolayer; (d) binding of HBsAg; (e) blocking with HRP.

2005a). More recently, an amperometric enzyme immunosensor with amplified sensitivity, based on layer-by-layer assembly of gold nanoparticles and thionine immobilized on a Nafion-modified electrode surface by electrostatic adsorption, has been reported for the detection of AFP. As in the configuration cited above, HRP was employed not only to block the possible remaining active sites of the nano-Au monolayer but also to amplify the response of the antigen–antibody reaction (Zhuo et al. 2006). Furthermore, mixed self-assembled monolayers, gold nanoparticles, and enzyme amplification were used together with microelectronic technology to develop an immunosensor for AFP based on microelectrodes and microwell systems constructed by SU-8 photoresist on silicon wafers (Xu et al. 2006).

Determination of the carcinoma antigen 125 (CA125) was carried out by using carbon electrodes modified with gold nanoparticles. An electrochemical immunosensor was designed by immobilizing anti-CA125 on a thionine and gold nanoparticles–modified carbon paste interface (Tang et al. 2006). The same antibody was also immobilized on colloidal gold nanoparticles to form a bioconjugate stabilized with a cellulose acetate membrane on the surface of a GCE. This configuration was proposed to develop a competitive immunoassay format to detect CA125 antigen with HRP-labeled CA125 antibody as tracer, and *o*-phenylenediamine and hydrogen peroxide as enzyme substrates (Wu et al. 2006).

Carbohydrate antigen 19-9 (CA19-9) is one of the most important carbohydrate tumor markers, expressed in many malignancies such as as pancreatic, colorectal, gastric, and hepatic carcinomas. An immunosensor for the rapid determination of CA19-9 in human serum has been developed by immobilization of the antibody in colloidal gold nanoparticle–modified carbon paste electrodes and monitoring of the direct electrochemistry of HRP labeled to a CA19-9 antibody. The formation of the antigen–antibody complex blocked the electron transfer of HRP toward the electrode substrate, which resulted in a significant current decrease (Dan et al. 2007). Another carbohydrate antigen, CA125, was detected using anti-CA125 gold hollow microspheres and porous polythionine–modified GCEs. The gold hollow microspheres greatly amplified the coverage of anti-CA125 molecules on the electrode surface. Electrochemical detection was accomplished by the amperometric changes occurring before and after the antigen–antibody interaction (Fu 2007).

4.1.2. Electrochemical Oxidation of Labeling Metal NPs

Another method for the determination of human IgG employed gold nanoparticle accumulation using magnetic particles (Chen et al. 2007). Goat anti-human IgG was immobilized on the wells of microtiter plates. The human IgG analyte was first captured by the primary antibody and then sandwiched by secondary

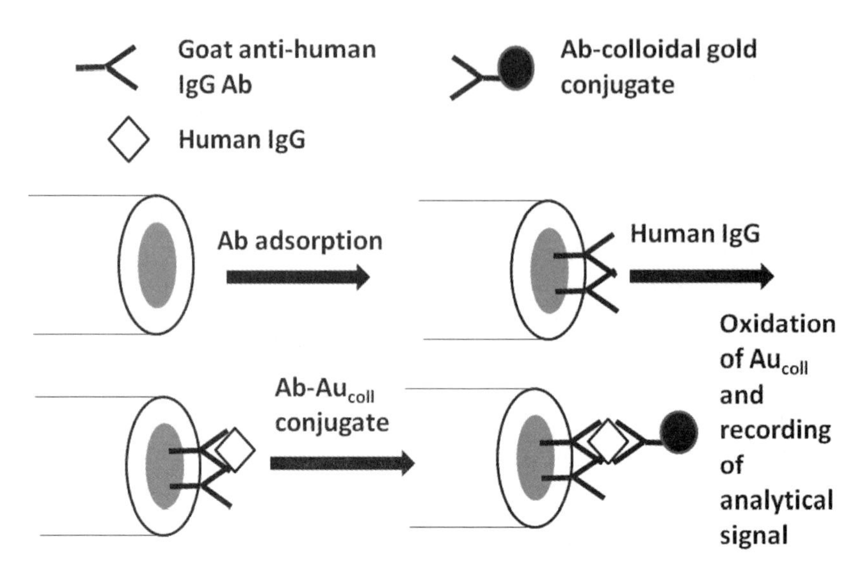

Figure 7.6. Scheme displaying the construction and the transduction principle of a sandwich immunosensor for human IgG using colloidal gold as electrochemical label.

antibody labeled with dethiobiotin. The conjugate of biotin-modified magnetic particles with captured avidin-functionalized gold nanoparticles and was reacted with dethiobiotinylated antibody followed by addition of biotin solution to wash it down, which caused the dissociation of the conjugates. The release of gold nanoparticles was then followed by anodic stripping voltammetry. This example shows that nanoparticles can be used not only as a medium to retain biomolecules with high stability, but also to provide versatile labels for the amplification of biosensing events. The secondary dissolution of the captured nanoparticles enables the amplified detection of the respective analyte by the release of many ions/molecules as a result of a single recognition event. This methodology was applied in a sensitive immunosensor for human IgG based on a sandwich-type assay using colloidal gold as electrochemical label (Chen et al. 2006a). As displayed in Figure 7.6, the capture protein was immobilized on a carbon paste electrode through passive adsorption to bind quantitatively with the corresponding antigen and colloidal gold-labeled antibody. In order to detect the amount of colloidal gold captured on the electrode surface, it was oxidized electrochemically to produce $AuCl_4^-$ ions, which were strongly adsorbed on the electrode surface and determined by adsorptive voltammetry.

Other configurations for IgG made use of silver precipitation on colloidal gold labels. After silver metal dissolution in an acidic solution, the antigen was indirectly determined by anodic stripping voltammetry at a GCE (Chu et al. 2005a).

A nanogold monolayer–modified chitosan-entrapped carbon paste electrode was used to construct an amperometric immunosensor for *Schistosoma japonicum*

(*Sj*) antigen. A sequential competitive immunoassay configuration was employed by loading of SjAb on a nanogold monolayer, then blocking in BSA solution, followed by competitive incubation in the buffer containing the SjAg analyte and SjAg labeled with HRP. Amperometric detection was made using hydroquinone as the enzyme substrate (Lei et al. 2003). A silver-enhanced colloidal gold metalloimmunoassay was also proposed for the determination of *Schistosoma japonicum* antibody in rabbit serum. The adult worm antigen SjAg was adsorbed on the walls of a polystyrene microwell and then reacted with the desired SjAb. Colloidal gold–labeled goat anti-rabbit IgG secondary antibody was adsorbed on the walls of the polystyrene microwells through the reaction with SjAb, followed by the silver enhancement process, dissolution of silver metal atoms in an acidic solution, and determination of dissolved silver ions by anodic stripping voltammetry at a GCE (Chu et al. 2005b).

4.1.3. Immunomagnetic Electrochemical Sensors

Electrochemical immunosensors are designed through the immobilization of the specific antibody on the surface of the electrochemical transducer. The main problem affecting immunosensors is reproducible regeneration of the sensing surface. The need for renewal of the sensing surface arises from the affinity constants derived from the strong antigen–antibody interaction. This renewal is a difficult task since the drastic procedures required alter immunoreagent bound to the surface of the transducer. This drawback makes immunosensors difficult to integrate into automatic systems. An alternative approach that avoids regeneration consists of using disposable antibody-coated magnetic nanoparticles and building up an *in situ* immunosensing surface by localizing the immunomagnetic NPs on the electrode area with the aid of a magnet. Moreover, the use of immunomagnetic NPs is particularly evident in the detection of analytes contained in complex sample matrices (e.g., heterogeneous food mixtures) that may exhibit either poor mass transport to immunosensor or physical blockage of immunosensor surface by nonspecific adsorption. The schematic representation of the enzyme-linked immunomagnetic electrochemical assay (ELIME) is presented in Figure 7.7 (Gehring et al. 1999). Immunogenic analyte (bacteria, for example) is sandwiched between an antibody-coated magnetic nanoparticle (immunomagnetic nanoparticle) and an antibody–enzyme conjugate. The immunomagnetic NP is trapped magnetically on the electrode surface, exposed to the enzymatic substrate, and the electroactive product is detected electrochemically. This type of immunomagnetic electrochemical assay was applied for different analytes with different transducer/enzyme combinations, gathered in Table 7.2. A good reproducibility of 2% relative standard deviation was observed (Santandreu et al. 1998).

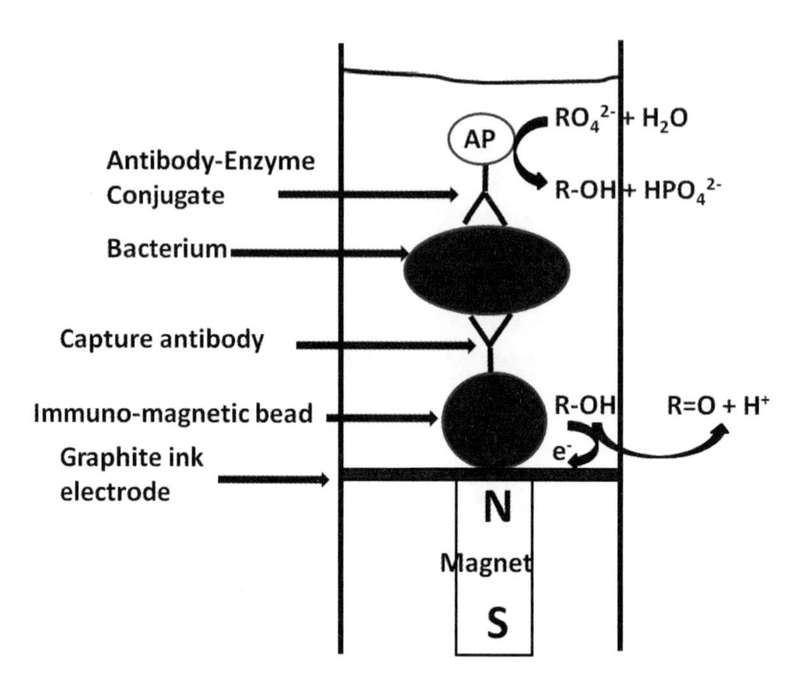

Figure 7.7. Schematic representation of the enzyme-linked immunomagnetic electrochemical assay (ELIME).

Table 7.2. Features of enzyme-linked immunomagnetic electrochemical assays using different enzymatic labels

Analyte	Transducer	Enzyme	Detection limit	Dynamic range	Ref.
Rabbit IgG	Graphite composite electrode	Horseradish peroxidase	9×10^{-6} μg/L	0–0.26 μM	Santandreu et al. 1998
E. coli	Graphite ink electrode	Alkaline phosphatase	4.7×10^{3} CFU/mL	0–10^{5} CFU/mL	Gehring et al. 1999
2,4-D herbicide	Nafion-SPE	Alkaline phosphatase	0.01 μg/L	0.01–100 μg/L	Dequaire et al. 1999
Human IgG	Carbon paste electrode	Horseradish peroxidase	0.18 μg/mL	0.51–30.17 μg/mL	Liu et al. 2006b

4.2. AuNP-BASED POTENTIOMETRIC SENSORS

A gold nanoparticle-based biomolecular immobilization method based on Nafion and gelatin was also described (Tang et al. 2004b). The detection of HBsAg was performed by measuring changes in the electric potential before and after the antigen–antibody reaction.

The self-assembling technique and opposite-charged adsorption methodology were combined for immobilization of diphtheria antibody, and applied to prepare an immunosensor for detecting diphtheria antigen. Anti-Diph was immobilized on nanometer-sized colloidal gold particles associated with polyvinyl butyral on a Pt electrode (PtE) (Tang et al. 2005a). Furthermore, a potentiometric immunosensor for direct and rapid detection of diphtherotoxin was developed by self-assembling of monoclonal diphtheria antibody onto a PtE based on the use of a gold nanoparticles–silica nanoparticles mixture and polyvinyl butyral as matrices. Anti-Diph was adsorbed onto the surface of the nanoparticles mixture, and then they were entrapped into polyvinyl butyral sol–gel network on a PtE. The immobilized DiphAb exhibited direct potentiometric response toward DiphAg. The immunosensor using the nanoparticles mixture exhibited much higher sensitivity, better reproducibility, and long-term stability than those constructed with gold nanoparticles or silica nanoparticles alone (Tang and Ren 2005b).

Carcinoembryonic antigen (CEA) is a well-known marker associated with progression of colorectal tumors. A current-amplified immunosensor for its determination was fabricated by coating negatively charged polysulfanilic acid–modified GCEs with positively charged toluidine blue. This approach provided an interface containing amine groups to assemble gold nanoparticles for immobilization of the carcinoembryonic antibody and HRP (Li et al. 2006). A dual-amplification strategy was proposed via backfilling gold nanoparticles on (3-mercaptopropyl) trimethoxysilane sol–gel (MPTS)–functionalized interface. MPTS acted as a building block for the electrode surface modification as well as a matrix for ligand functionalization with first amplification. The second signal amplification strategy was based on the backfilling immobilization of nanogold particles to the immunosensor surface. Using a noncompetitive design, membrane potential change occurred before and after the antigen–antibody interaction (An et al 2007). Electrochemical detection of prostate-specific antigen has been carried out by using a colloidal gold/alumina-derived sol–gel film (Liu 2008). An anti-PSA antibody-functionalized colloidal gold/alumina sol–gel film was prepared and PSA detection was accomplished on the basis of the potential change occurring before and after the antigen–antibody interaction.

4.3. IMPEDIMETRIC SENSORS

4.3.1. AuNP-Based Impedimetric Sensors

Hepatitis B virus surface antigen was detected using electrochemical impedance spectroscopy (EIS) through immobilization of the antibody onto gold nanoparticles–modified 4-aminothiophenol self-assembled monolayers (Wang et al. 2004a) or on the surface of a PtE modified with colloidal gold and polyvinyl butyral (Tang et al. 2004a).

Many examples of the determination of human IgG can be found in the literature. For example, a capacitive sensing method using gold nanoparticles self-assembled to a sol–gel–modified electrode was developed for the direct detection of the human IgG in serum. Because the mercaptopropyltriethoxysilane film that was formed was ultrathin, the immobilization density of antibodies was high because of the high surface–volume ratio of the assembled gold nanoparticles. The capacitive immunosensor provided high sensitivity, and no cross-reactivity was observed with other proteins (Wu et al. 2005).

A highly sensitive electrochemical impedance immunosensor was also developed using an amplification procedure with a colloidal gold–labeled antibody as the primary amplifying probe, and a multistep amplification sequence. Rabbit anti-human IgG antibody was immobilized through a self-assembled colloidal gold layer on a gold electrode. The analyte, human IgG, was detected through impedance measurements with the sensing interface modified by the rabbit anti-human IgG antibody. In the primary amplification, the colloidal gold–labeled goat anti-human IgG antibody was used to amplify the electron transfer resistance resulting from the antigen binding to the immunosensor surface. A further amplification was performed through sequential binding of the colloidal gold–labeled rabbit anti-goat IgG and the colloidal gold–labeled goat anti-human IgG antibodies (Chen et al. 2006a).

An electrochemical immunoassay for human chorionic gonadotrophin (hCG) was proposed using a conductive colloidal gold nanoparticle/titania sol–gel composite membrane deposited on a GCE via a vapor deposition method. Horseradish peroxidase–labeled hCG antibody (HRP-anti-hCG) was encapsulated into the composite architecture. Similar to that commented on above, the formation of the immunoconjugate between hCG and the immobilized HRP-anti-hCG introduced a barrier for the direct electrical communication between the immobilized enzyme and the electrode surface that could be monitorized by EIS (Chen et al. 2006b).

Impedance sensing was also used to detect allergen–antibody interaction on a GCE modified by gold electrodeposition (Huang et al. 2007). In this application, allergen Der f2 was immobilized through gold nanocrystals deposited on the electrode surface, the charge transfer kinetics of the $[Fe(CN)_6]^{3-/4-}$ redox pair being monitored. The allergen–antibody interactions that occurred on the electrode surface altered the interfacial electron transfer resistance, R_{CT}, by preventing the redox species approaching the electrode. The results showed that R_{CT} increased with increasing concentration of monoclonal antibodies.

4.3.2. CNT-Based Impedimetric Immunosensors

Measurement of D-dimer has become an essential element in the diagnosis of deep-vein thrombosis and pulmonary embolism; in this context, microelectrodes with an area of 9×10^{-4} cm^2 were used to develop an impedimetric immunosensor for detecting deep venous thrombosis biomarker (D-dimer). The biosensor is based on functionalized carbon nanotubes (SWCNT-COOH) where the antibody (anti-D-dimer) was immobilized by covalent binding. At low frequencies, the impedance modulus $|Z(f)|$ decreases clearly with an increase of D-dimer concentration, which indicates that a larger amount of D-dimer interacts with the biofunctionalized surface. This reaction leads to a decrease in the electron transfer resistance. The increase of CNT conductivity can be explained by the increase of the negative charge of the immune complex, which would induce an increase of CNT conductivity by a field effect: increase of the density of positive free carriers in CNTs (Maehashi et al. 2007).

An impedimetric microimmunosensor allows obtaining a sensitivity of 40.1 kmM^{-1} and a detection limit of 0.1 pg/mL (0.53 fM) with a linear range from 0.1 pg/mL to 2 mg/mL (0.53 fM to 0.01 mM). Using carbon nanotubes and microelectrodes, high sensitivity and dynamic range were obtained. The biosensor exhibited a short response time of 10 min. Moreover, the studied immunosensor exhibited good reproducibility (R.S.D. 8.2%, $n = 4$) (Bourigua et al. 2010).

4.3.3. Immunomagnetic Impedancemetric Sensors

Electrochemical impedance measurements of the electrical properties of an antibody layer immobilized on a gold electrode allows direct monitoring of the variation of these properties when antigen–antibody interaction occurs. This technique allows label-free detection of the antigen concentration in biological samples. The problem of regeneration of the sensing surface has been solved, in this example, by using streptavidin magnetic microbeads for the immobilization of the antibody specific of a small pesticide molecule, the atrazine. The antibody, biotinyl-anti-atrazine Fab fragment K47, forms a quite stable layer on the streptavidin-magnetic microbeads immobilized on a gold electrode using a 300-mT magnet, due to the high affinity of the biotin–streptavidin interaction. After the antibody layer formation an antigen, atrazine was injected and interacted with the antibody (Helali et al. 2006).

In order to obtain a calibration data set, the values of electron-transfer resistance differences, ΔR_{et}, versus the added atrazine concentrations were plotted. The change of electron transfer resistance was calculated according to Eq. (7.5):

$$\Delta R_m = R_{et(Ab)} - R_{et(Ab\text{-}Ag)} \qquad (7.5)$$

where $R_{et(Ab)}$ is the value of electron transfer resistance after antibody immobilization and $R_{et(Ab\text{-}Ag)}$ is the value of the electron transfer resistance after antigen binding to the antibody. A linear relationship between ΔR_{et} values and the concentration of atrazine was established in the range from 50 to 500 ng/ml. A detection limit of 10 ng/ml was reached (Helali et al. 2006).

A label-free immunosensor for the detection of ochratoxin A (OTA) based on use of magnetic nanoparticles (MNPs) was developed. A gold electrode was modified using bovine serum albumin conjugate with a glutaraldehyde–thiolamine linker, creating a layer that prevents nonspecific binding of OTA on gold. The OTA antibodies were attached to MNPs using the carbodiimide chemistry and afterwards were immobilized on the modified gold electrode using a strong magnetic field. The impedance variation due to the specific antibody–OTA interaction was correlated with the OTA concentration in the samples. The increase in electron-transfer resistance values was proportional to the concentration of OTA in a linear range between 0.01 and 5 ng/mL, with a detection limit of 0.01 ng/mL. SPR measurements showed a larger response range (1–50 ng/mL) with a detection limit of 0.94 ng/mL. Analytical results were in accordance with a standard ELISA test kit (Zamfir et al. 2011).

An immunomagnetic biosensor for the label-free detection of a bacterial model, *Escherichia coli,* has been described and compared to a self assembled multilayer system. The paramagnetic nanoparticles layer attracted to, and formed on, the gold electrode surface via a magnetic field up to 300 mT is not totally blocking for the redox probe compared to the thiol self-assembled monolayer (a biotin thiol and a spacer thioalcohol). Moreover, the modeling of the Nyquist spectra obtained by electrochemical impedance spectroscopy for increasing concentrations of *E. coli* show for both systems a sigmoid variation of the polarization resistance with increasing logarithmic concentration of bacteria. A sensitivity slope of 10.675 was obtained for the immunomagnetic sensor, compared to 6.832 for the self-assembled multilayer process, indicating the higher sensitivity of the paramagnetic nanoparticle biosensor (Maalouf et al. 2008).

4.4. CONDUCTOMETRIC SENSORS

4.4.1. AuNP-Based Conductometric Immunosensors

Aflatoxin B1 (AFB1) was determined using an electrode fabricated by self-assembling HRP and AFB1 antibody onto gold nanoparticles. These provided a suitable microenvironment for the immobilized biomolecules and decreased the electron transfer impedance. The formation of the antibody–antigen complex by a simple one-step immunoreaction introduced a barrier for the direct electrical communication between the immobilized HRP and the electrode surface. Therefore, local conductivity variations could be detected, giving good conductometric response relative to AFB1 (Liu et al. 2006a).

4.4.2. MNP-Based Conductometric Immunosensors

A new approach has been reported for immunoassays based on magnetic nanoparticles for *Escherichia coli* detection using conductometric measurements. Biotinylated antibodies, anti-*E. coli*, were immobilized on streptavidin-modified magnetite nanoparticles by biotin–streptavidin interaction. A layer of functionalized nanoparticles were directly immobilized on the conductometric electrode using glutaraldehyde cross-linking.

The specific test with *E. coli* cells was investigated by conductometric measurements. Results show good response as a function of antigen additions. The specific detection of 1 CFU/ml of *E. coli* induces a conductivity variation of 35 μS. A negative test on *Staphylococcus epidermidis* showed no conductometric response. Conductometric measurements allowed the detection of 500 CFU/L of *E. coli* (Hnaiein et al. 2008).

5. CONCEPTION AND MODELING OF AMPLIFICATION EFFECT IN NANOMATERIAL-BASED DNA BIOSENSORS

5.1. AMPEROMETRIC SENSORS

Electrochemical DNA biosensors constitute useful analytical tools for sequence-specific DNA diagnosis and detection due to their inherent advantages of low cost, sensitivity, and rapidity of response (Odenthal and Gooding 2007). Similarly to other electrochemical biosensors, gold nanoparticles can play an important role both in DNA immobilization on electrode surfaces and as suitable labels to improve detection of hybridization events.

Electrode surface-immobilization techniques for the achievement of stable and highly dense single-stranded DNA (ssDNA) monolayers are a key aspect in the development of DNA biosensors. In this context, gold nanoparticle films provide a suitable means for the ssDNA immobilization. For example, self-assembly of colloidal Au onto a gold electrode resulted in easier attachment of an oligonucleotide with a mercaptohexyl group at the 5′-phosphate end and, therefore, an increased capacity for nucleic acid detection. The ssDNA surface density on the colloidal Au-modified gold electrode was 1.0×10^{14} molecules/cm^2, ca. 10 times larger than on a bare gold electrode (Cai et al. 2001).

5.1.1. Electrochemical Oxidation of Metal NPs

In order to achieve adequate sensitivity, the development of novel amplification methodologies for quantitative DNA sensing events is essential. Gold nanoparticles can play a crucial role to this respect. DNA hybridization detection was accomplished by electrochemical stripping of the colloidal gold tag. In this protocol, the hybridization of a target oligonucleotide to magnetic bead-linked oligonucleotide probes was followed by binding of streptavidin-coated metal nanoparticles to the captured DNA, dissolution of the nanometer-sized gold tag, and potentiometric stripping measurements of the dissolved metal tag at single-use thick-film carbon electrodes. An advanced magnetic processing technique was used to isolate the DNA duplex and to provide low-volume mixing. Further signal amplification, and lowering of the detection limits to the nanomolar and picomolar domains was achieved by precipitating gold or silver, respectively, onto the colloidal gold label. This electrochemical stripping metallogenomagnetic approach coupled the inherent signal amplification of stripping metal analysis with discrimination against nonhybridized DNA, the use of microliter sample volumes, and disposable transducers (Wang et al. 2001a). A scheme of the sensor preparation and functioning is displayed in Figure 7.8.

The use of gold nanoparticle–DNA probes and subsequent signal amplification by silver enhancement constitutes a somewhat similar approach. The assay relied on the electrostatic adsorption of target oligonucleotides onto the sensing surface of a GCE and its hybridization to the gold nanoparticle–labeled oligonucleotides

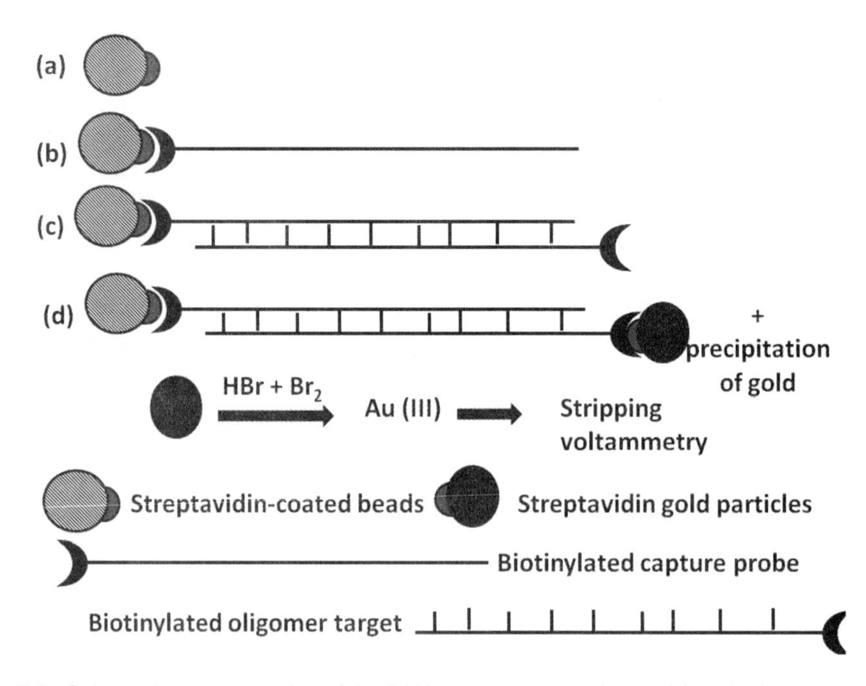

Figure 7.8. Schematic representation of the DNA sensor preparation and functioning.

DNA probe. After silver deposition onto gold nanoparticles, binding events between probe and target were monitored by the DPV signal of the large number of silver atoms anchored on the hybrids at the electrode surface. The signal intensity difference permitted distinguishing between the match of two perfectly matched DNA strands and the near-perfect match where just one base pair was wrong. Due to this "nanoparticle-promoted" silver signal amplification, a detection limit of 50 pM of complementary oligonucleotides could be obtained (Cai et al. 2002).

A genosensor based on the modification of a pencil graphite electrode with target DNA and further hybridization with complementary probes conjugated to Au nanoparticles was developed for the detection of the Factor V Leiden mutation from polymerase chain reaction (PCR)–amplified real samples. Specific probes were immobilized onto the Au nanoparticles in two different modes: (1) inosine-substituted probes were covalently attached from their amino groups at the 5′ end using *N*-(3-dimethylamino)propyl)-*N*-ethylcarbodiimide hydrochloride and *N*-hydroxysulfosuccinimide onto a carboxylate-terminated L-cysteine self-assembled monolayer preformed on the Au nanoparticles; (2) probes with a hexanethiol group at their 5′-phosphate end formed a SAM on Au nanoparticles. The appearance of the Au oxidation signal at +1.20 V was employed to detect hybridization. Discrimination between the homozygous and heterozygous mutations was established by comparing the peak currents of the Au signals. The detection limit for the PCR amplicons was found to be as low as 0.78 fmol (Ozsoz et al. 2003).

Electrochemical genosensors offer a promising alternative to carry out applications directed toward gene analysis, detection of genetic disorders, tissue matching, and forensic applications, due to their high sensitivity, small dimensions, low cost, and compatibility with microfabrication technology. Furthermore, increasing information about the human genome requires fast and simple methods for the detection of single-nucleotide polymorphisms (SNPs). SNPs can be also coded by monitoring the changes in the electrochemical signal of monobase-modified colloidal gold nanoparticles. A chitosan layer formed on the alkanethiol SAM-modified Au nanoparticle can be used to attach the monobases through their 5′-phosphate group via the formation of a phosphoramidate bond with the free amino groups of chitosan. If there is a SNP in DNA and the mismatched bases are complementary to the monobase, Au nanoparticles accumulate on the electrode surface in the presence of DNA polymerase I (Klenow fragment), thus resulting in a significant change in the Au oxide wave. Monobase-modified Au nanoparticles showed not only the presence of a SNP, but also identified which bases were involved within the pair. A model study was performed by using a synthetic 21-base DNA probe related to tumor necrosis factor (TNF-r) along with all its possible mutant combinations (Kerman et al. 2004a). Figure 7.9 shows how the bases involved in a SNP can be identified by comparing the voltammetric signals obtained from the four different monobase-modified gold nanoparticles.

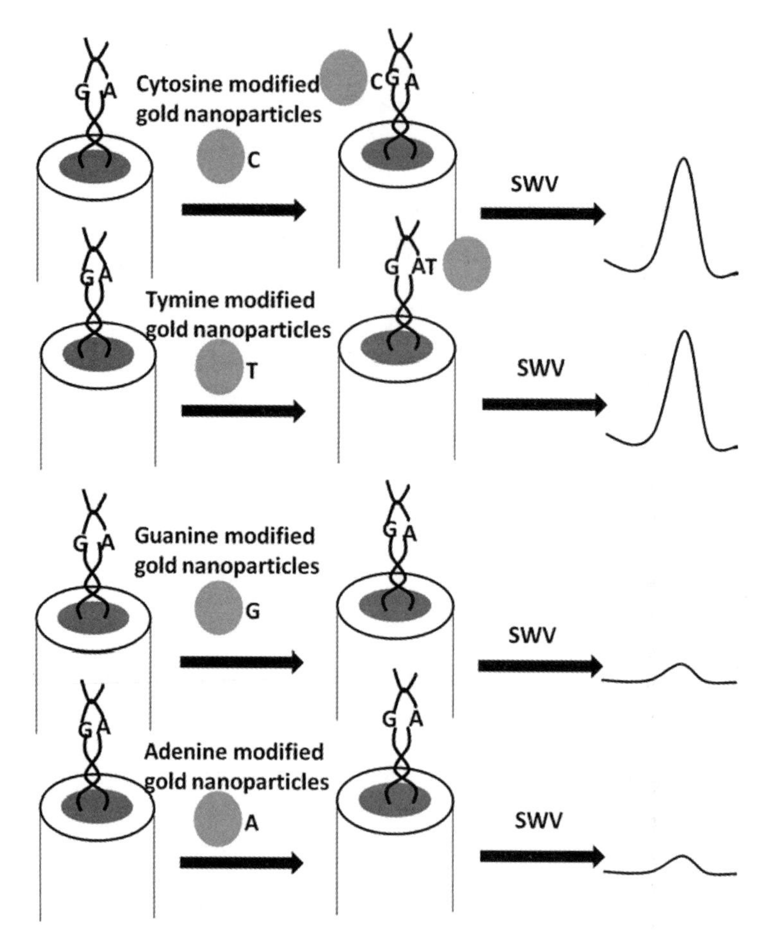

Figure 7.9. Scheme displaying voltammetric identification of the bases involved in a SNP using monobase-modified Au nanoparticles.

5.1.2. CNT-Based Electrochemical DNA Sensors

DNA biosensors based on nucleic acid recognition processes are rapidly being developed toward the goal of rapid, simple, and inexpensive testing for genetic and infectious diseases. Electrochemical hybridization biosensors rely on the immobilization of a single-stranded (ss-) DNA probe onto the transducer surface that converts the duplex formation into a useful electrical signal (Gooding 2002). The performance of such devices can benefit greatly from the use of carbon nanotubes. Such improvements are attributed to enhanced detection of the target guanine or of the product of an enzyme label, as well as to the use of CNT carrier platforms.

Surface-confined multiwall carbon nanotubes have been shown to be useful to facilitate the adsorptive accumulation of the guanine nucleobase and greatly enhances its oxidation signal (Wang et al. 2003a). This advantage of CNT-coated glassy carbon electrodes has been illustrated from comparison with the common unmodified glassy carbon, carbon paste, and graphite pencil electrodes. The dramatic amplification of the guanine signal has been combined with a label-free electrical detection of DNA hybridization. A similar enhancement of the guanine DNA response was reported at MWCNT paste electrodes (Pedano and Rivas 2004) and at a SWCNT-coated glassy carbon electrode (Wang et al. 2004b). Trace amounts (mg/L) of nucleic acids have thus been detected following short accumulation times.

An array of vertically aligned MWCNTs, embedded in SiO_2, has been shown to be useful for ultrasensitive detection of DNA hybridization (Li et al. 2003). Subattomoles of DNA targets have been measured by combining the CNT nanoelectrode array with $Ru(bpy)_3^{2+}$-mediated guanine oxidation. Such a CNT array was also applied for label-free detection of DNA PCR amplicons, and offered the detection of less than 1000 target amplicons (Lin et al. 2004). A lower nanotube density offered higher sensitivity. Improved sensitivity of electrical DNA hybridization has been reported also in connection with the use of daunomycin redox intercalator (Cai et al. 2003a). A MWCNT-COOH modified glassy carbon was used for this task along with a 5′-amino group–functionalized oligonucleotide probe and pulse-voltammetric transduction.

5.1.3. Genomagnetic Electrochemical Assays

Currently developed DNA hybridization sensors are using single-stranded (ss) short (15–25 nucleotides) oligodeoxynucleotides (ODN, probe DNA) immobilized on an electrode. The ODN-modified electrode is immersed in the target DNA solution. When the sequence of target DNA matches that of the probe (based complementary Watson-Crick pairing), a probe–target (hybrid) duplex DNA is formed at the electrode surface. The hybridization event (DNA duplex formation) is detected electrochemically in various ways. This system works quite well with synthetic ODNs when probe target DNAs are about the same lengths. In a real DNA sequence analysis with longer PCR products, viral or chromosomal DNAs, the target DNAs are substantially longer than the probe. With longer target DNAs, difficulties connected with nonspecific target DNA adsorptions frequently arise, resulting in a loss of specificity and decreased sensitivity. Elimination of the nonspecific DNA adsorption at the electrodes (such as carbon or gold ones), usually applied for DNA hybridization, is very difficult. A new method based on separation of the DNA hybridizing step (on magnetic microbeads) from the electrochemical detection step has been proposed and successfully used (Palacek et al. 2002;

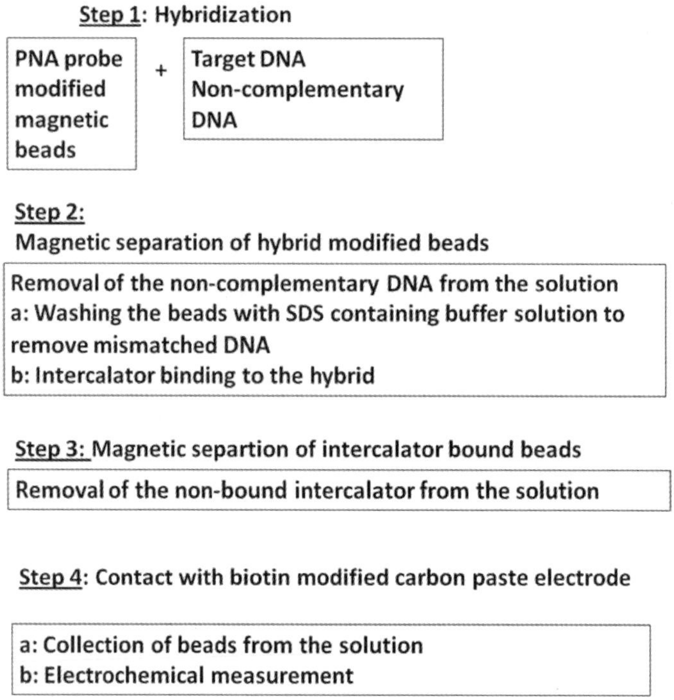

Step 1: Hybridization

PNA probe modified magnetic beads + Target DNA Non-complementary DNA

Step 2:
Magnetic separation of hybrid modified beads

Removal of the non-complementary DNA from the solution
a: Washing the beads with SDS containing buffer solution to remove mismatched DNA
b: Intercalator binding to the hybrid

Step 3: Magnetic separtion of intercalator bound beads

Removal of the non-bound intercalator from the solution

Step 4: Contact with biotin modified carbon paste electrode

a: Collection of beads from the solution
b: Electrochemical measurement

Figure 7.10. Procedure of the genomagnetic bioassay.

Table 7.3. Features of the different genomagnetic electrochemical assays

Analyte	Electrochemical transducer	Label	Detection limit	Dynamic range	Ref.
Breast cancer BRCA1 gene	Potentiometric stripping measurement Graphite pencil electrode	No	100 ng/mL (ppb)	100 ppb–20 ppm	Wang et al. 2001b
Breast cancer BRCA1 gene	Differential pulse voltammetry	Alkaline phosphatase	100 ppb	0.25–1.0 ppm (20-min hybridization)	Wang et al. 2002
21-mer DNA	Square-wave voltammetry	Meldol's blue intercalator	2 ppb	2–20 pM	Kerman et al. 2004b
DNA sequence from *Salmonella*	Differential pulse voltammetry Graphite composite electrodes	No	9.68 pM		Erdem et al. 2006

Wang et al. 2001b). One procedure of the genomagnetic electrochemical bioassay (GEME), based on a PNA probe and an electroactive intercalator, is schematized in Figure 7.10 (Kerman et al 2004b).

The detection of target DNA can be done by label-free detection of guanine peak (Wang et al. 2001; Palacek et al. 2002; Erdem et al. 2006), by detection of chemically modified DNA prelabeled or postlabeled by redox probes (Palacek et al. 2002), by detection of the product of the enzymatic reaction on DNA postlabeled by enzyme (Wang et al. 2002), or by detection of redox intercalators (Kerman et al. 2004). The features of the different genomagnetic electrochemical assays are presented in Table 7.3.

5.2. IMPEDIMETRIC SENSORS

5.2.1. AuNP-Based DNA Biosensors

The security of food made from transgenic plants is a worldwide matter of concern. An important associated aspect is accurate, sensitive, and rapid detection of the transgenic plants. DNA electrochemical sensors are most likely to become an analytical tool for transgenic plant products. Recently, a DNA electrochemical biosensor was described for electrochemical impedance spectroscopy detection of the sequence-specific DNA related to PAT transgene in the transgenic plants. Poly-2,6-pyridinedicarboxylic acid film (PDC) was fabricated by electropolymerizing 2,6-pyridinedicarboxylic acid on a glassy carbon electrode. Gold nanoparticles were attached on the PDC/GCE, and then probe DNA was immobilized by the interaction of gold nanoparticles with DNA. The electron transfer resistance of the electrode surface in the $[Fe(CN)_6]^{3-/4-}$ solution increased after the immobilization of the DNA probe on the nAu/PDC/GCE. The hybridization of the DNA probe with complementary DNA made electron transfer resistance increase further. The presence of gold nanoparticles on the PDC dramatically enhanced the amount of immobilized probe DNA and greatly improved the sensitivity of DNA detection. The difference between the electron-transfer resistance value at the ssDNA/nAu/PDC/GCE and at the dsDNA/nAu/PDC/GCE was used as the signal for detecting the PAT gene fragment, with a detection limit of 2.4×10^{-11} M (Yang et al. 2007).

Another interesting research area is the preparation of DNA-modified electrodes to study the interaction of DNA with other molecules. For example, a simple, stable, and repeatable approach based on the preparation of a DNA-modified indium tin oxide (ITO) electrode was used for the determination of mifepristone, which is a kind of progesterone receptor compound used mainly as a contraceptive that can be intercalated into a DNA double helix. The electrode consisted of (3-aminopropyl)trimethoxysilane, gold nanoparticles, and DNA molecules self-assembled to the ITO electrode surface (Xu et al. 2001). Furthermore, the analysis of the interfacial interaction between DNA-binding drugs, used mostly as antibiotics and in cancer therapy, with immobilized DNA probes has a particular role in the rational design of new DNA-binding drugs and drug screening. These interfacial interactions can be investigated by impedance spectroscopy. Using gold nanoparticle–deposited substrates, impedance spectroscopy resulted in a 20- to 40-fold increase in the detection limit (Wang et al. 2004b). Arrays of deposited gold nanoparticles on gold electrodes offered a convenient tool to subtly control probe immobilization to ensure suitably adsorbed DNA orientation and accessibility of other binding molecules (Li et al. 2005).

5.2.2. CNT-Based DNA Biosensors

Cai et al. (2003b) described indicator-free AC impedance measurements of DNA hybridization based on DNA probe–doped polypyrrole film over a MWCNT layer. The hybridization event led to a decrease of impedance values reflecting the reduction of the electrode resistance. A fivefold sensitivity enhancement was observed compared to analogous measurements without CNTs.

5.2.3. MNP-Based DNA Biosensors

A new approach based on DNA hybridization for detecting HBV virus using non-Faradaic electrochemical impedance spectroscopy has been reported (Hassen et al. 2008). DNA probes modified with biotin in the 5′ position were immobilized on streptavidin-modified magnetic nanoparticles by biotin–streptavidin interaction. A layer of functionalized MNPs was immobilized on directly a bare gold electrode using a magnet. Hybridization reactions with a specific complementary DNA target and with no complementary target were investigated using non-Faradaic impedance spectroscopy. Results show good immobilization of the DNA probes and hybridization with different concentrations of complementary DNA. Non-Faradaic impedance spectroscopy allows the detection of 50 pmol of HBV DNA and 160 pmol of HIV DNA on a sample of 20 μL. Saturations were reached for the same concentration, 12.65 nmol/mL, for the same quantity of immobilized DNA probes.

6. CONCLUSION

The intensive work carried out recently on the development of nanomaterials-based biosensors shows clearly the potentialities and advantageous features of this approach to construct biosensors exhibiting enhanced performance with respect to previous designs. The unique properties of nanomaterials concerning immobilization of biomolecules retaining their biological activity makes them a powerful tool to modify electrode materials and to construct robust and sensitive biosensors which can find application in many fields of interest.

REFERENCES

Agui L., Manso J., Yanez-Sedeno P., and Pingarro J.M. (2006) Amperometric biosensor for hypoxanthine based on immobilized xanthine oxidase on nanocrystal gold–carbon paste electrodes. *Sens. Actuators B* **113**, 272–280. DOI: 10.1016/j.snb.2005.03.001

Aguilar-Arteaga K., Rodriguez J.A., and Barrado E. (2010) Magnetic solids in analytical chemistry: A review. *Anal. Chim. Acta* **674**, 157–165. DOI: 10.1016/j.aca.2010.06.043

Albers J., Grunwald T., Nebling E., Piechtta G., and Hintsche R. (2003) Electrical biochip technology—A tool for microarrays and continuous monitoring. *Anal. Bioanal. Chem.* **377**, 521–527. DOI: 10.1007/s00216-003-2192-7

An H.Z., Yuan R.T., Tang D.P., Chai Y., and Li N. (2007) Dual-amplification of antigen–antibody interactions via backfilling gold nanoparticles on (3-mercaptopropyl) trimethoxysilane sol-gel functionalized interface. *Electroanalysis* **19**, 479–486. DOI: 10.1002/elan.200603752

Andreescu S., Njagi J., Ispas C., and Ravalli M.T. (2009) JEM spotlight: Applications of advanced nanomaterials for environmental monitoring. *J. Environ. Monit.* **11**, 27–40. DOI: 10.1039/b811063h

Baron R., Zayat M., and Willner I. (2005) Dopamine-, L-DOPA-, adrenaline-, and noradrenaline-induced growth of Au nanoparticles: Assays for the detection of neurotransmitters and of tyrosinase activity. *Anal. Chem.* **77**, 1566–1571. DOI: 10.1021/ac048691v

Baughman R.H., Zakhidov A., and de Heer W.A. (2002) Carbon nanotubes—The route toward applications. *Science* **297**, 787–792. DOI: 10.1126/science.1060928

Besteman K., Lee J., Wiertz F., Heering H., and Dekker C. (2003) Enzyme-coated carbon nanotubes as single-molecule biosensors. *Nano Lett.* **3**, 727–730. DOI: 10.1021/nl034139u

Bourdillon C., Demaille C., Gueris J., Moiroux J., and Saveant J.M. (1993) A fully active monolayer enzyme electrode derivatized by antigen-antibody attachment. *J. Am. Chem. Soc.* **115**, 12264–12269. DOI: 10.1021/ja00079a005

Bourigua S., Hnaien M., Bessueille F., Lagarde F., Dzyadevych S., Maaref A., Bausells J., Errachid A., and Jaffrezic-Renault N. (2010) Impedimetric immunosensor based on SWCNTCOOH modified gold microelectrodes for label-free detection of deep venous thrombosis biomarker. *Biosens. Bioelectron.* **26**, 1278–1282. DOI: 10.1016/j.bios.2010.07.004

Bourigua S., Maaref A., Bessueille F., and Jaffrezic Renault N. (2013) A new design of electrochemical and optical biosensor based on biocatalytic growth of Au nanoparticles—Example of glucose detection. *Electroanalysis* **25**, 644–651. DOI: 10.1002/elan.201200243

Britto P.J., Santhanam K.S.V., and Ayajan P.M. (1996) Carbon nanotube electrode for oxidation of dopamine. *Bioelectrochem. Bioenerg.* **41**, 121–125. DOI: 10.1016/0302-4598(96)05078-7.

Cai H., Xu C., He P., and Fang Y. (2001) Colloid Au-enhanced DNA immobilization for the electrochemical detection of sequence-specific DNA. *J. Electroanal. Chem.* **510**, 78–85. DOI: 10.1016/S0022-0728(01)00548-4

Cai H., Wang Y., He P., and Fang Y. (2002) Electrochemical detection of DNA hybridization based on silver-enhanced gold nanoparticle label. *Anal. Chim. Acta* **469**, 165–172. DOI: 10.1016/S0003-2670(02)00670-0

Cai H., Cao X., Jiang Y, He P., and Fang Y. (2003a) Carbon nanotube-enhanced electrochemical DNA biosensor for DNA hybridization detection. *Anal. Bioanal. Chem.* **375**, 287–293. DOI: 10.1007/s00216-002-1652-9

Cai H., Xu Y., He P., and Fang Y.Z. (2003b) Indicator free DNA hybridization detection by impedance measurement based on the DNA-doped conducting polymer film formed on the carbon nanotube modified electrode. *Electroanalysis* **15**, 1864–1870. DOI: 10.1002/elan.200302755

Carralero Sanz C., Mena M.L., Gonzalez-Cortes A., Yanez-Sedeno P., and Pingarron J.M. (2005) Development of a tyrosinase biosensor based on gold nanoparticles-modified glassy carbon electrodes: Application to the measurement of a bioelectrochemical polyphenols index in wines. *Anal. Chim. Acta* **528**, 1–8. DOI: 10.1016/j.aca.2004.10.007

Carralero C., Mena M.L., Gonzalez-Cortes A., Yanez-Sedeno P., and Pingarron J.M. (2006) Development of a high analytical performance-tyrosinase biosensor based on a composite graphite–Teflon electrode modified with gold nanoparticles. *Biosens. Bioelectron.* **22**, 730–736. DOI: 10.1016/j.bios.2006.02.012

Chen H., Jiang J.-H., Huang Y., Deng T., Li J.S., Li G., and Yu R.Q. (2006a) An electrochemical impedance immunosensor with signal amplification based on Au-colloid labeled antibody complex. *Sens. Actuators B* **117**, 211–218. DOI: 10.1016/j.snb.2005.11.026

Chen J., Tang J., Yan F., and Ju H. (2006b) A gold nanoparticles/sol–gel composite architecture for encapsulation of immunoconjugate for reagentless electrochemical immunoassay. *Biomaterials* **27**, 2313–2321. DOI: 10.1016/j.biomaterials.2005.11.004

Chen Z., Peng Z., Zhang P., Jin X., Jiang J., Zhang X., Shen G., and Yu R. (2007) A sensitive immunosensor using colloidal gold as electrochemical label. *Talanta* **72**, 1800–1804. DOI: 10.1016/j.talanta.2007.02.020

Chu X., Fu X., Chen K., Shen G.L., and Yu R.Q. (2005a) An electrochemical stripping metalloimmunoassay based on silver-enhanced gold nanoparticle label. *Biosens. Bioelectron.* **20**, 1805–1812. DOI: 10.1016/j.bios.2004.07.012

Chu X., Xiang Z.F., Fu X., Wang S.P., Shen G.L., and Yu R.Q. (2005b) Silver-enhanced colloidal gold metalloimmunoassay for *Schistosoma japonicum* antibody detection. *J. Immunol. Methods* **301**, 77–88. DOI: 10.1016/j.jim.2005.03.005

Cubukcu M., Timur S., and Anik U. (2007) Examination of performance of glassy carbon paste electrode modified with gold nanoparticle and xanthine oxidase for xanthine and hypoxanthine detection. *Talanta* **74**, 434–439. DOI: 10.1016/j.talanta.2007.07.039

Dan D., Xiaoxing X., Shengfu W., and Zhang A. (2007) Reagentless amperometric carbohydrate antigen 19-9 immunosensor based on direct electrochemistry of immobilized horseradish peroxidase. *Talanta* **71**, 1257–1262. DOI: 10.1016/j.talanta.2006.06.028

Davis J.J., Coleman K., Azamian B., Bagshaw C., and Green M.L. (2003) Chemical and biochemical sensing with modified single walled carbon nanotubes. *Chem. Eur. J.* **9**, 3732–3739. DOI: 10.1002/chem.200304872

Dequaire M., Degrand C., and Limoges B. (1999) An immunomagnetic electrochemical sensor based on a perfluorosulfonate-coated screen-printed electrode for the determination of 2,4-dichlorooxyacetic acid. *Anal. Chem.* **71**, 2571–2577. DOI: 10.1021/ac990101j

Dolatabadi J.E.N., Mashinchian O., Ayoubi B., Jamali A. A., Mobed A., Losic D., Omidi Y., and de la Guardia M. (2011) Optical and electrochemical DNA nanobiosensors. *Trends Anal. Chem.* **30**, 459–472. DOI: 10.1016/j.trac.2010.11.010

Elyacoubi A., Zayed S., Blankert B., and Kauffmann J-M. (2006) Development of an amperometric enzymatic biosensor based on gold modified magnetic nanoporous microparticles, *Electroanalysis* **18**, 345–350. DOI: 10.1002/elan.200503418

Erdem A., Pividori M.I., Lermo A., Bonanni A., Del Valle M., and Alegret S. (2006) Genomagnetic assay based on label-free electrochemical detection using magneto-composite electrodes. *Sens. Actuators B* **114**, 591--598. DOI: 10.1016/j.snb.2005.05.031

Fu X.H. (2007) Electrochemical immunoassay for carbohydrate antigen-125 based on polythionine and gold hollow microspheres modified glassy carbon electrodes. *Electroanalysis* **19**, 1831–1839. DOI: 10.1002/elan.200703943

Gehring A.G., Brewster J.D., Irwin P.L., Tu S.I., and Van Houten L.J. (1999) 1-Naphtyl phosphate as an enzymatic substrate for enzyme-linked immunomagnetic electrochemistry. *J. Electroanal. Chem.* **469**, 27–33. DOI: 10.1016/S0022-0728(99)00183-7

Gooding J.J. (2002) Electrochemical DNA hybridization biosensors. *Electroanalysis* **14**, 1149–1156. DOI: 10.1002/1521-4109(200209)14:17<1149::AID-ELAN1149>3.0.CO;2-8

Gooding J.J., Wibowo R., Liu J.Q., Yang W., Losic D., Orbons S., Mearns F.J., Shapter J.G., and Hibbert D.B. (2003) Protein electrochemistry using aligned carbon nanotube arrays. *J. Am. Chem. Soc.* **125**, 9006–9007. DOI: 10.1021/ja035722f.

Gooding J.J. and Hibbert D.B. (1999) The application of alkanethiol self-assembled monolayers to enzyme electrodes. *Trends Anal. Chem.* **18**, 525–533. DOI: 10.1016/S0165-9936(99)00133-8

Gorton L. (1995) Carbon paste electrodes modified with enzymes, tissues, and cells. *Electroanalysis* **7**, 23–45. DOI: 10.1002/elan.1140070104

Guiseppi-Elie A., Lei C., and Baughman R. (2002) Direct electron transfer of glucose oxidase on carbon nanotubes. *Nanotechnology* **13**, 559–564. DOI: 10.1088/0957-4484/13/5/303

Hassen W.M., Chaix C., Abdelghani A., Bessueille F., Leonard D., and Jaffrezic-Renault N. (2008) An impedimetric DNA sensor based on functionalized magnetic nanoparticles for HIV and HBV detection. *Sens. Actuators B* **134**, 755–760. DOI: 10.1016/j.snb.2008.06.020

Helali S., Martelet C., Abdelghani A., Maaref M.A., and Jaffrezic-Renault N.A (2006) Disposable immunomagnetic electrochemical sensor based on functionalised magnetic beads on gold surface for the detection of atrazine. *Electrochim. Acta* **51**(24), 5182–5186. DOI: 10.1016/j.electacta.2006.03.086

Heller A. (1990) Electrical wiring of redox enzymes. *Acc. Chem. Res.* **23**, 128–134. DOI: 10.1021/ar00173a002

Hirsch A. (2002) Functionalization of single-walled carbon nanotubes. *Angew. Chem. Int. Ed.* **41**, 1853–1859. DOI: 10.1002/1521-3773(20020603)41:11<1853::AID-ANIE1853>3.0.CO;2-N

Hnaiein M., Hassen W.M., Abdelghani A., Fournier-Wirth C., Coste J., Bessueille F., Leonard D., and Jaffrezic-Renault N. (2008) A conductometric immunosensor based on functionalized magnetite nanoparticles for *E. coli* detection. *Electrochem. Commun.* **10**, 1152–1154. DOI: 10.1016/j.elecom.2008.04.009

Hrapovic S., Liu Y., Male K., and Luong J.H. (2004) Electrochemical biosensing platforms using platinum nanoparticles and carbon nanotubes. *Anal. Chem.* **76**, 1083–1088. DOI: 10.1021/ac035143t

Hsing I. M., Xu Y., and Zhao W. (2007) Micro- and nano-magnetic particles for applications in biosensing. *Electroanalysis* **19**, 755–768. DOI: 10.1002/elan.200603785

Huang H., Ran P., and Liu Z. (2007) Impedance sensing of allergen–antibody interaction on glassy carbon electrode modified by gold electrodeposition. *Bioelectrochemistry* **70**, 257–262. DOI: 10.1016/j.bioelechem.2006.10.002

Hwang S., Kim E., and Kwak J. (2005) Electrochemical detection of DNA hybridization using biometallization. *Anal. Chem.* **77**, 579–584. DOI: 10.1021/ac048778g

Jdanova A.S., Poyard S., Soldatkin A.P., Jaffrezic-Renault N., and Martelet C. (1996) Conductometric urea sensor. Use of additional membranes for the improvement of its analytical characteristics. *Anal. Chim. Acta* **321**, 35–40. DOI: 10.1016/0003-2670(95)00548-X

Katz E., Willner I., and Wang J. (2004) Electroanalytical and bioelectroanalytical systems based on metal and semiconductor nanoparticles. *Electroanalysis* **16**, 19–44. DOI: 10.1002/elan.200302930

Kauffmann J.-M., Yu D., El Yacoubi A., and Blankert B. (2006) Magnetic nanoporous microparticles for biosensors and bioreactors. *LabPlus Int.* **20**(3), 6–8.

Kerman K., Matsubara Y., Morita Y., and Takamura Y. (2004) Peptide nucleic acid modified magnetic beads for intercalator based electrochemical detection of DNA hybridization. *Sci. Technol. Adv. Mater.* **5**, 351–357. DOI: 10.1016/j.stam.2004.01.009

Kerman K., Saito M., Morita Y., Takamura Y., Ozsoz M., and Tamiya E. (2004) Electrochemical coding of single-nucleotide polymorphisms by monobase-modified gold nanoparticles. *Anal. Chem.* **76**, 1877–1884. DOI: 10.1021/ac0351872

Koehne J., Chen H., Li J., Cassell A., Ye Q., Ng H., and Meyyappan M. (2003) Ultrasensitive label-free DNA analysis using an electronic chip based on carbon nanotube nanoelectrode arrays. *Nanotechnology* **14**, 1239–1245. DOI: 10.1088/0957-4484/14/12/001

Kuramitz H. (2009) Magnetic microbead-based electrochemical immunoassays. *Anal. Bioanal. Chem.* **394**, 61–69. DOI: 10.1007/s00216-009-2650-y

Lei C.X., Gong F.C., Shen G.L., and Yu R.Q. (2003) Amperometric immunosensor for *Schistosoma japonicum* antigen using antibodies loaded on a nano-Au monolayer modified chitosan-entrapped carbon paste electrode. *Sens. Actuators B* **96**, 582–588. DOI: 10.1016/j.snb.2003.06.001

Li C.Z., Liu Y., and Luong J.H.T. (2005) Impedance sensing of DNA binding drugs using gold substrates modified with gold nanoparticles. *Anal. Chem.* **77**, 478–485. DOI: 10.1021/ac048672l

Li J., Ng H.T., Cassell A., Fan W., Chen H., Ye Q., Koehne J., and Meyyappan M. (2003) Carbon nanotube nanoelectrode array for ultrasensitive DNA detection. *Nano Lett.* **3**, 597–602. DOI: 10.1021/nl0340677

Li X., Yuan R., Chai Y., Zhang L., Zhuo Y., and Zhang Y. (2006) Amperometric immunosensor based on toluidine blue/nano-Au through electrostatic interaction for determination of carcinoembryonic antigen. *J. Biotechnol.* **123**, 356–366. DOI: 10.1016/j.jbiotec.2005.11.023

Liao M.-H., Guo J.-C., and Chen W.C. (2006) A disposable amperometric ethanol biosensor based on screen-printed carbon electrodes mediated with ferricyanide-magnetic nanoparticle mixture. *J. Magnetism Magnetic Mater.* **304**(1), e421–e423. DOI: 10.1016/j.jmmm.2006.01.223

Liao S.Z., Qiao Y.A., Han W.T., Xie Z.X., Wu Z.Y., Shen G.L., and Yu R.Q. (2012) Acetylcholinesterase liquid crystal biosensor based on modulated growth of gold nanoparticles for amplified detection of acetylcholine and inhibitor. *Anal. Chem.* **84**, 45–49. DOI: 10.1021/ac202895j

Lim S.Y., Lee J.S., and Park C.B. (2010) In situ growth of gold nanoparticles by enzymatic glucose oxidation within alginate gel matrix. *Biotechnol. Bioeng.* **105**, 210–214. DOI: 10.1002/bit.22519

Lin Y., Lu F., Tu Y., and Ren Z. (2004) Glucose biosensors based on carbon nanotube nanoelectrode ensembles. *Nano Lett.* **2**, 191–195. DOI: 10.1021/nl0347233

Liu S.Q. and Ju H.X. (2002) Renewable reagentless hydrogen peroxide sensor based on direct electron transfer of horseradish peroxidase immobilized on colloidal gold-modified electrode. *Anal. Biochem.* **307**, 110–116. DOI: 10.1016/S0003-2697(02)00014-3

Liu S., Leech D., and Ju H. (2003a) Application of colloidal gold in protein immobilization, electron transfer, and biosensing. *Anal. Lett.* **36**, 1–19. DOI: 10.1081/AL-120017740

Liu S., Yu J., and Ju H. (2003b) Renewable phenol biosensor based on a tyrosinase-colloidal gold modified carbon paste electrode. *J. Electroanal. Chem.* **540**, 61–67. DOI: 10.1016/S0022-0728(02)01276-7

Liu S.Q. and Ju H.X. (2003) Reagentless glucose biosensor based on direct electron transfer of glucose oxidase immobilized on colloidal gold modified carbon paste electrode. *Biosens. Bioelectron.* **19**, 177–183. DOI: 10.1016/S0956-5663(03)00172-6

Liu Y. (2008) Electrochemical detection of prostate-specific antigen based on gold colloids/alumina derived sol-gel film. *Thin Solid Film* **516**, 1803–1808. DOI: 10.1016/j.tsf.2007.08.048

Liu Y., Qin Z., Wu X., and Jiang H. (2006a) Immune-biosensor for aflatoxin B_1 based bioelectrocatalytic reaction on micro-comb electrode. *Biochem. Eng. J.* **32**, 211–217. DOI: 10.1016/j.bej.2006.10.003

Liu Z., Liu Y., Yang H., Yang Y., Shen G., and Yu R. (2005) A phenol biosensor based on immobilizing tyrosinase to modified core-shell magnetic nanoparticles supported at a carbon paste electrode. *Anal. Chim. Acta* **533**(1), 3–9. DOI: 10.1016/j.aca.2004.10.077

Liu Z.M., Yang H.F., Li Y.F., Liu Y.L., Shen G.L., and Yu R.Q. (2006b) Core-shell magnetic nanoparticles applied for immobilization of antibody on carbon paste electrode and amperometric immunosensing. *Sens. Actuators B* **113**, 956–962. DOI: 10.1016/j.snb.2005.04.002

Lu B.-W. and Chen W.C. (2006) A disposable glucose biosensor based on drop-coating of screen-printed carbon electrodes with magnetic nanoparticles. *J. Magnetism Magnetic Mater.* **304**(1), e400–e402. DOI: 10.1016/j.jmmm.2006.01.222

Luong J.H., Hrapovic S., Wang D., Bensebaa F., and Simard B. (2004) Solubilization of multiwall carbon nanotubes by 3-aminopropyltriethoxysilane towards the fabrication of electrochemical biosensors with promoted electron transfer. *Electroanalysis* **16**, 132–139. DOI: 10.1002/elan.200302931

Maalouf R., Hassen W.M. Fournier-Wirth C., Coste J., and Jaffrezic-Renault N. (2008) Comparison of two innovatives approaches for bacterial detection: Paramagnetic nanoparticles and self-assembled multilayer processes. *Microchim. Acta*, **163**, 157–161. DOI: 10.1007/s00604-008-0008-3

Maehashi K., Katsura T., Kerman K., Takamura Y., Matsumoto K., and Tamiya E. (2007) Label-free protein biosensor based on aptamer-modified carbon nanotube field-effect transistors. *Anal. Chem.* **79**, 782–787. DOI: 10.1021/ac060830g

Marcus R.A., and Sutin N. (1985) Electron transfers in chemistry and biology. *Biochim. Biophys. Acta* **811**, 265–322. DOI: 10.1016/0304-4173(85)90014-X

Mena M.L., Yanez-Sedeno P., and Pingarron J.M. (2005) A comparison of different strategies for the construction of amperometric enzyme biosensors using gold nanoparticle-modified electrodes. *Anal. Biochem.* **336**, 20–27. DOI: 10.1016/j.ab.2004.07.038

Miyabayashi A. and Mattiasson B. (1988) An enzyme electrode based on electromagnetic entrapment of the biocatalyst bound to magnetic beads. *Anal. Chim. Acta* **213**, 121–130. DOI: 10.1016/S0003-2670(00)81346-X

Musameh M., Wang J., Merkoci A., and Lin Y. (2002) Low-potential stable NADH detection at carbon-nanotube-modified glassy carbon electrodes. *Electrochem. Commun.* **4**, 743–746. DOI: 10.1016/S1388-2481(02)00451-4

Nouira W., Maaref A., Elaissari A., Vocanson F., Siadat M., and Jaffrezic-Renault N. (2013) Comparative study of conductometric glucose biosensor based on gold and on magnetic nanoparticles. *Mater. Sci. Eng. C* **33**, 298–303. DOI: 10.1016/j.msec.2012.08.043

Nouira W., Maaref A., Vocanson F., Siadat M., Saulnier J., Lagarde F., and Jaffrezic-Renault N. (2012) Enhancement of enzymatic IDE biosensor response using gold nanoparticles. Example of the detection of urea. *Electroanalysis* **24**(5), 1088–1094. DOI: 10.1002/elan.201100681

Odenthal K.J. and Gooding J.J. (2007) An introduction to electrochemical DNA biosensors. *Analyst* **132**, 603–610. DOI: 10.1039/b701816a.

Ozsoz M., Erdem A., Kerman K., Ozkan D., Tugrul B., Topcuoglu N., Ekren H., and Taylan M. (2003) Electrochemical genosensor based on colloidal gold nanoparticles for the detection of factor V Leiden mutation using disposable pencil graphite electrodes. *Anal. Chem.* **75**, 2181–2187. DOI: 10.1021/ac026212r

Palacek E., Fojta M., and Jelen F. (2002) New approaches in the developement of DNA sensors: Hybridization and electrochemical detection of DNA and RNA at two different surfaces. *Bioelectrochemistry* **56**, 85–90. DOI: 10.1016/S1567-5394(02)00025-7

Palecek E. and Fojta M. (2007) Magnetic beads as versatile tools for electrochemical DNA and protein biosensing. *Talanta* **74**, 276–290. DOI: 10.1016/j.talanta.2007.08.020

Pänke O., Balkenhohl T., Kafka J., Schäfer and D., Lisdat F. (2008) Impedance spectroscopy and biosensing. *Adv. Biochem. Eng. Biotechnol.* **109**, 195–237. DOI: 10.1007/10_2007_081

Patolsky F., Weizmann Y., and Willner I. (2004) Long-range electrical contacting of redox enzymes by SWCNT connectors. *Angew. Chem. Int. Ed.* **43**, 2113–2117. DOI: 10.1002/anie.200353275

Pavlov V., Xiao Y., and Willner I. (2005) Inhibition of the acetycholine esterase-stimulated growth of Au nanoparticles: Nanotechnology-based sensing of nerve gases. *Nano Lett.* **5**, 649–653. DOI: 10.1021/nl050054c

Pedano M. and Rivas G.A. (2004) Adsorption and electrooxidation of nucleic acids at carbon nanotubes paste electrodes. *Electrochem. Commun.* **6**, 10–16. DOI: 10.1016/j.elecom.2003.10.008

Rao C.N., Satishkumar B.C., Govindaraj A., and Nath M. (2001) Nanotubes. *ChemPhysChem.* **2**(2), 78–105. DOI: 10.1002/1439-7641(20010216)2:2<78::AID-CPHC78>3.0.CO;2-7

Rashid M.H., Bhattacharjee R.R., Kota A., and Mandal T.K. (2006) Synthesis of spongy gold nanocrystals with pronounced catalytic activities. *Langmuir* **22**, 7141–7143. DOI: 10.1021/la060939j

Rubianes M.D. and Rivas G.A. (2003) Carbon nanotubes paste electrode. *Electrochem. Commun.* **5**, 689–694. DOI: 10.1016/S1388-2481(03)00168-1

Santandreu M., Sole S., Fabregas E., and Alegret S. (1998) Development of electrochemical immunosensing systems with renewable surfaces *Biosens. Bioelectron.* **13**, 7–17. DOI: 10.1016/S0956-5663(97)00096-1

Sheppard N.F. Jr., Mears D.J., and Guiseppi-Elie A. (1996) Model of an immobilized enzyme conductimetric urea biosensor. *Biosens. Bioelectron.* **11**, 967–979. DOI: 10.1016/0956-5663(96)87656-1

Shulga O. and Kirchhoff J.R. (2007) An acetylcholinesterase enzyme electrode stabilized by an electrodeposited gold nanoparticle layer. *Electrochem. Commun.* **9**, 935–940. DOI: 10.1016/j.elecom.2006.11.021

Simon de Dios A., and Diaz-Garcia M.E. (2010) Multifunctional nanoparticles: Analytical prospects. *Anal. Chim. Acta* **666**, 1–22. DOI: 10.1016/j.aca.2010.03.038

Siqueira J.R. Jr., Abouzar M.H., Poghossian A., Zucolotto V., Oliveira O.N. Jr., and Schöning M.J. (2009a) Penicillin biosensor based on a capacitive field-effect structure functionalized with a dendrimer/carbon nanotube multilayer. *Biosens. Bioelectron.* **25**, 497–501. DOI: 10.1016/j.bios.2009.07.007

Siqueira J.R. Jr., Werner C.F., Bäcker M., Poghossian A., Zucolotto V., Oliveira O.N. Jr., and Schöning M.J. (2009a) Layer-by-layer assembly of carbon nanotubes incorporated in light-addressable potentiometric sensors. *J. Phys. Chem. C* **113**, 14765–14770. DOI: 10.1021/jp904777t

Solé S., Merkoci A., and Alegret S. (2001) New materials for electrochemical sensing III. Beads. *Trends Anal. Chem.* **20**(2), 102–110. DOI: 10.1016/S0165-9936(00)00059-5

Sotiropoulou S. and Chaniotakis N.A. (2003) Carbon nanotube array-based biosensor. *Anal. Bioanal. Chem.* **375**, 103. DOI: 10.1007/s00216-002-1617-z

Tang D., Yuan R., Chai Y., Dai J., Zhong X., and Liu Y. (2004a) A novel immunosensor based on immobilization of hepatitis B surface antibody on platinum electrode modified colloidal gold and polyvinyl butyral as matrices *via* electrochemical impedance spectroscopy. *Bioelectrochemistry* **65**, 15–22. DOI: 10.1016/j.bioelechem.2004.05.004

Tang D., Yuan R., Chai Y.Q., Zhong X., Liu Y., Dai J, and Zhang L.Y. (2004b) Novel potentiometric immunosensor for hepatitis B surface antigen using a gold nanoparticle-based biomolecular immobilization method. *Anal. Biochem.* **333**, 345–350. DOI: 10.1016/j.ab.2004.06.035

Tang D., Yuan R., Chai Y., Zhang L., Zhong X., Liu Y., and Dai J. (2005a) Preparation and application on a kind of immobilization method of anti-diphtheria for potentiometric immunosensor modified colloidal Au and polyvinyl butyral as matrixes. *Sens. Actuators B* **104**, 199–206. DOI: 10.1016/j.snb.2004.04.116

Tang D. and Ren J.J. (2005b) Direct and rapid detection of diphtherotoxin via potentiometric immunosensor based on nanoparticles mixture and polyvinyl butyral as matrixes. *Electroanalysis* **17**, 2208–2216. DOI: 10.1002/elan.200503351

Tang D., Yuan R., and Chai Y. (2006) Electrochemical immuno-bioanalysis for carcinoma antigen 125 based on thionine and gold nanoparticles-modified carbon paste interface. *Anal. Chim. Acta* **654**, 158–165. DOI: 10.1016/j.aca.2006.01.094

Teja A.S. and Koh P.Y. (2009) Synthesis, properties, and applications of magnetic iron oxide nanoparticles. *Prog. Cryst. Growth Charact. Mater.* **55**, 22–45. DOI: 10.1016/j.pcrysgrow.2008.08.003

Temple-Boyer P., BenYahia A., Sant W., Pourciel-Gouzy M.L., Launay J., and Martinez A. (2008) Modelling of urea-EnFETs for haemodialysis applications. *Sens. Actuators B* **131**, 525–532. DOI: 10.1016/j.snb.2007.12.037

Terrill R.H., Postlethwaite T.A., Chen C.H., Poon C.D., Terzis A., Chen A., Hutchison J.E., Clark M.R., Wignall G., Londono J.D., Superfine R., Falvo M., Johnson C.S. Jr., Samulski E.T., and Murray R.W. (1995) Monolayers in three dimensions: NMR, SAXS, thermal, and electron hopping studies of alkanethiol stabilized gold clusters. *J. Am. Chem. Soc.* **117**, 12537–12548. DOI: 10.1021/ja00155a017

Valentini F., Amine A., Orlanducci S., Terranova M., and Palleschi G. (2003) Carbon nanotube purification: Preparation and characterization of carbon nanotube paste electrodes. *Anal. Chem.* **75**, 5413–5421. DOI: 10.1021/ac0300237

Varlan A.R., Sansen W., Van Loey A., and Hendrickx M. (1996) Covalent enzyme immobilization on paramagnetic polyacrolein beads. *Biosens. Bioelectron.* **11**(4), 443–448. DOI: 10.1016/0956-5663(96)82740-0

Varlan A.R., Suls J., Jacobs P., and Sansen W. (1995) A new technique of enzyme entrapment for planar biosensors. *Biosens. Bioelectron.* **10**(8), XV–XIX. DOI: 10.1016/0956-5663(95)96967-4

Wang J., Xu D., Kawde A.N., and Polsky R. (2001a) Metal nanoparticle-based electrochemical stripping potentiometric detection of DNA hybridization. *Anal. Chem.* **73**, 5576–5581. DOI: 10.1021/ac0107148

Wang J., Kawde A.N., Erdem A., and Salazar M. (2001b) Magnetic bead-based label-free electrochemical detection of DNA hybridization. *Analyst* **126**, 2020–2024. DOI: 10.1039/b106343j

Wang J., Xu D., Erdem A., Polsky R., and Salazar M.A. (2002) Genomagnetic electrochemical assays of DNA hybridization. *Talanta* **56**, 931–938. DOI: 10.1016/S0039-9140(01)00653-1

Wang J., Kawde A., and Mustafa M. (2003a) Carbon-nanotube-modified glassy carbon electrodes for amplified label-free electrochemical detection of DNA hybridization. *Analyst* **128**, 912–916. DOI: 10.1039/b303282e

Wang J., Musameh M., and Lin Y. (2003b) Solubilization of carbon nanotubes by Nafion toward the preparation of amperometric biosensors. *J. Am. Chem. Soc.* **125**, 2408–2409. DOI: 10.1021/ja028951v

Wang J. and Musameh M. (2003a) Enzyme-dispersed carbon-nanotube electrodes: A needle microsensor for monitoring glucose. *Analyst* **128**, 1382–1385. DOI: 10.1039/b309928h

Wang J. and Musameh M. (2003b) Carbon nanotube/teflon composite electrochemical sensors and biosensors. *Anal. Chem.* **75**, 2075–2079. DOI: 10.1021/ac030007+

Wang M., Wang L., Wang G., Ji X., Bai Y., Li T., Gong S., and Li J. (2004a) Application of impedance spectroscopy for monitoring colloid Au-enhanced antibody immobilization and antibody–antigen reactions. *Biosens. Bioelectron.* **19**, 575–582. DOI: 10.1016/S0956-5663(03)00252-5

Wang J., Li M., Shi Z., Li N., and Gu Z. (2004b) Electrochemistry of DNA at single-wall carbon nanotubes. *Electroanalysis* **16**, 140–144. DOI: 10.1002/elan.200302915

Wang J. and Musameh M. (2004) Carbon nanotube screen-printed electrochemical sensors. *Analyst* **129**, 1–2. DOI: 10.1039/b313431h

Wang J. (2005) Carbon-nanotube based electrochemical biosensors: A review. *Electroanalysis* **17**(1), 7–14. DOI: 10.1002/elan.200403113

Willner I. (2002) Biomaterials for sensors, fuel cells, and circuitry. *Science* **298**, 2407–2408. DOI: 10.1126/science.298.5602.2407

Willner I., Willner B., and Tel-Vered R. (2011) Electroanalytical applications of metallic nanoparticles and supramolecular nanostructures. *Electroanalysis* **23**, 13–28. DOI: 10.1002/elan.201000506

Wu L., Chen J., Du D., and Ju H. (2006) Electrochemical immunoassay for CA125 based on cellulose acetate stabilized antigen/colloidal gold nanoparticles membrane. *Electrochim. Acta* **51**, 1208–1214. DOI: 10.1016/j.electacta.2005.06.011

Wu Z.S., Li J.S., Luo M.H., Shen G.L., and Yu R.Q. (2005) A novel capacitive immunosensor based on gold colloid monolayers associated with a sol–gel matrix. *Anal. Chim. Acta* **528**, 235–242. DOI: 10.1016/j.aca.2004.09.075

Xiao Y., Patolsky F., Katz E., Hainfeld J.F., and Willner I. (2003) "Plugging into enzymes": Nanowiring of redox enzymes by a gold nanoparticle. *Science* **299**, 1877–1881. DOI: 10.1126/science.1080664

Xiao Y., Pavlov V., Levine S., Niazov T., Markovitch G., and Willner I. (2004) Catalytic growth of Au nanoparticles by NAD(P)H cofactors: Optical sensors for $NAD(P)^+$-dependent biocatalyzed transformations. *Angew. Chem. Int. Ed.* **43**, 4519–4522. DOI: 10.1002/anie.200460608

Xiao Y., Pavlov V., Shlyahovsky B., and Willner I. (2005) An Os^{II}–bisbipyridine–4-picolinic acid complex mediates the biocatalytic growth of Au nanoparticles: Optical detection of glucose and acetylcholine esterase inhibition. *Chem. Eur. J.* **11**, 2698–2704. DOI: 10.1002/chem.200400988

Xu J., Zhu J.J., Zhu Y., Gu K., and Chen H.Y. (2001) A novel biosensor of DNA immobilization on nano-gold modified ITO for the determination of mifepristone. *Anal. Lett.* **34**, 503–512. DOI: 10.1081/AL-100002591

Xu Y.Y., Bian C., Chen S., and Xia S. (2006) A microelectronic technology based amperometric immunosensor for α-fetoprotein using mixed self-assembled monolayers and gold nanoparticles *Anal. Chim. Acta* **561**, 48–54. DOI: 10.1016/j.aca.2005.12.061

Yan Y.M., Tel-Vered R., Yehez Keli O., Cheglakov Z., and Willner I. (2008) Biocatalytic growth of Au nanoparticles immobilized on glucose oxidase enhances the ferrocene-mediated bioelectrocatalytic oxidation of glucose. *Adv. Mater.* **20**, 2365–2370. DOI: 10.1002/adma.200703128

Yanez-Sedeno P. and Pingarron J.M. (2005) Gold nanoparticle-based electrochemical biosensors. *Anal. Bioanal. Chem.* **382**, 884–886. DOI: 10.1007/s00216-005-3221-5

Yang J., Yang T., Feng Y., and Jiao K. (2007) A DNA electrochemical sensor based on nanogold-modified poly-2,6-pyridinedicarboxylic acid film and detection of PAT gene fragment. *Anal. Biochem.* **365**, 24–30. DOI: 10.1016/j.ab.2006.12.039

Ye J., Wen Y., Zhang W.D., Gan L.M., Xu G., and Sheu F.S. (2004) Nonenzymatic glucose detection using multi-walled carbon nanotube electrodes. *Electrochem. Commun.* **6**, 66–70. DOI: 10.1016/j.elecom.2003.10.013

Yu D., Blankert B., Bodoki E., Bollo S., Viré J.-C., Sandulescu R., Nomura A., and Kauffmann J.-M. (2006) Amperometric biosensor based on horseradish peroxidase-immobilised magnetic microparticles. *Sens. Actuators B* **113**, 749–754. DOI: 10.1016/j.snb.2005.07.026

Yu X., Chattopadhyay D., Galeska I., Papadimitrakopoulos F., and Rusling J.F. (2003) Peroxidase activity of enzymes bound to the ends of single-wall carbon nanotube forest electrodes. *Electrochem. Commun.* **5**, 408–411. DOI: 10.1016/S1388-2481(03)00076-6

Zamfir L.G., Geana I., Bourigua S., Rotariu L., Bala C., Errachid A., and Jaffrezic-Renault N. (2011) Highly sensitive label-free immunosensor for ochratoxin A based on functionalized magnetic nanoparticles and EIS/SPR detection. *Sens. Actuators B* **159**, 178–184. DOI: 10.1016/j.snb.2011.06.069

Zayats M., Baron R., Popov I., and Willner I. (2005) Biocatalytic growth of Au nanoparticles: From mechanistic aspects to biosensors design. *Nano Lett.* **5**, 21–25. DOI: 10.1021/nl048547p

Zhang H.F., Liu R.X., and Sheng Q.L. (2011) Enzymatic deposition of Au nanoparticles on the designed electrode surface and its application in glucose detection. *Coll. Surf.* **82**, 532–535. DOI: 10.1016/j.colsurfb.2010.10.012

Zhang S., Wang N., Yu H., Niu Y., and Sun C. (2005) Covalent attachment of glucose oxidase to an Au electrode modified with gold nanoparticles for use as glucose biosensor. *Bioelectrochemistry* **67**, 15–22. DOI: 10.1016/j.bioelechem.2004.12.002

Zhao Q., Gan Z., and Zhuang Q. (2002) Electrochemical sensors based on carbon nanotubes. *Electroanalysis* **14**, 1609–1613. DOI: 10.1002/elan.200290000

Zhuo Y., Yuan R., Chai Y., Tang D., Zhang Y., Wang N., Li W., and Zhu Q. (2005a) A reagentless amperometric immunosensor based on gold nanoparticles/thionine/Nafion-membrane-modified gold electrode for determination of α-1-fetoprotein. *Electrochem. Commun.* **7**, 355–360. DOI: 10.1016/j.elecom.2005.02.001

Zhuo Y., Yuan R., Chai Y., Zhang Y., Li X., Wang N., and Zhu O. (2006) Amperometric enzyme immunosensors based on layer-by-layer assembly of gold nanoparticles and thionine on Nafion modified electrode surface for α-1-fetoprotein determinations. *Sens. Actuators B* **114**, 631–639. DOI: 10.1016/j.snb.2005.04.051

Zhuo Y., Yuan R., Chai Y., Zhang Y., Wang N., Li X., Zhu Q., and Wang N. (2005b) An amperometric immunosensor based on immobilization of hepatitis B surface antibody on gold electrode modified gold nanoparticles and horseradish peroxidase. *Anal. Chim. Acta* **548**, 205–210. DOI: 10.1016/j.aca.2005.05.058

CHAPTER 8

Ion-Sensitive Field-Effect Transistors with Nanostructured Channels and Nanoparticle-Modified Gate Surfaces

Theory, Modeling, and Analysis

V. K. Khanna

1. INTRODUCTION

The basic concept of the ion-sensitive field-effect transistor (ISFET) was introduced more than four decades ago by P. Bergveld (2003), who modified the conventional metal-oxide-semiconductor field-effect transistor (MOSFET) by discarding the gate metal, and replacing it with an ion-selective membrane, an electrolyte, and a reference electrode. As a result of this modification, the device could detect and measure the concentration of particular ionic species with a proton or pH sensor as the starting point. These sensors are referred to as pH ISFETs and membrane ISFETs (MEMFETs), and fall under the umbrella of chemical FETs or CHEMFETs (Schöning and Poghossian 2006, 2008; Lee et al. 2009; Lee and Cui 2010). Subsequently, the device was used in the form of an enzyme FET, an immunological FET, or a DNA FET by immobilizing an enzyme, antibody, or DNA

DOI: 10.5643/9781606505984/ch8

strands on the gate dielectric. The generalized terms "BioFET" and "biosensor" were coined (Heitzinger et al. 2008). The biosensor utilizes the gating effect of the charged biomolecules on the carrier concentration in the semiconductor, and functions by modulating its conductance according to the charges. It has receptors on the oxide/dielectric surface of the semiconductor, which attract and bind to targets in the analyte solution. The charges on the target molecules influence the semiconductor, altering its conductance.

With the advent of nanotechnology, ideas were proposed for using nanostructured materials such as nanoporous silicon (Zehfroosh et al. 2010), carbon nanotubes (CNTs) (Dong et al. 2009, 2012), and silicon nanowires (Si NWs) (Fan and Lu 2006; Lee and Shin 2009; Lin et al. 2011; Reddy et al. 2011; Vieira et al. 2012) in ISFET fabrication. Here, the focus of attention was the ISFET channel. As clinical and biological interests in ISFETs are growing, and biomolecules have nanoscale dimensions, it was thought that the microscopic ISFET channel must be dimensionally reduced to harness nanoscale effects.

A vital consequence of miniaturization to the nanoscale is the breakdown of macroscopic properties accompanied by the emergence of new quantum-level phenomena and related effects. Although the obvious advantages of nano effects have been experimentally corroborated by several research publications, very few papers have dwelt on the underlying sensing mechanisms. An in-depth understanding of the fundamental sensing principles, design considerations, optimization criteria to achieve the desired response characteristics, optimal operational regimes for maximum sensitivity, and ultimate detection limits is provided through physico-chemical and computational models of nano-ISFET biosensors (Ushaa and Eswaran 2012). The primary motivation for this chapter is the need to address these issues with regard to ISFET-based nanoscale biosensors (Figure 8.1). A closely allied corollary discipline explores the use of nanoparticles in interfacing the biomolecules to the gate of a conventional ISFET to enhance the response (Katz et al. 2004; Luo et al. 2006; Kim et al. 2011). This subject of using nanoparticle formulations to elicit higher sensitivity from traditional microchannel ISFETs will also form a part of this chapter. The chapter is organized as follows. Section 2 outlines the new nano-ISFET technologies. Section 3 describes the basic physical principles of Si nanowire biosensors. Salient features and outcomes of the Nair-Alam model of silicon nanowire biosensors are discussed in Section 4. In Section 5, the pH-sensing behavior of nano-ISFETs is elucidated. Section 6 highlights the optimal-sensitivity operational domain of the device. The reason for the high apparent sensitivity values attained in dual-gated nanowire ISFETs, beyond the theoretical boundary laid down by thermodynamic restrictions, is explained in Section 7 in terms of a capacitance model considering the coupling between the liquid and back gates. The proposal of a tunnel FET as a biosensor reported in recent literature (Sarkar and Banerjee 2012) is then briefly examined in Section 8. From Section 9 onward, attention diverts from nanostructuring ISFET channels

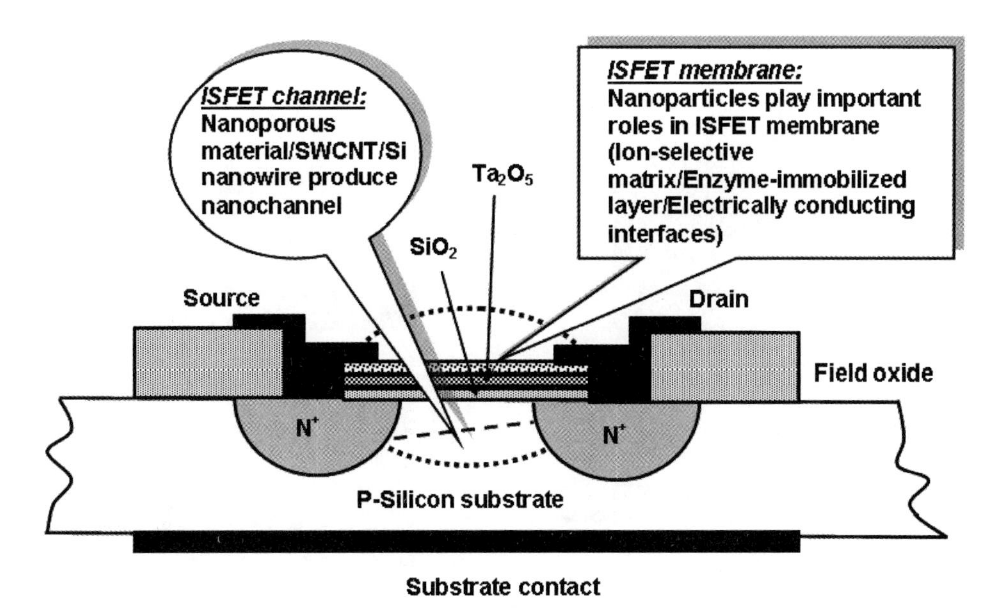

Figure 8.1. Enclosed by dotted line is the region of interest for nanostructuring. The effects of this nanostructuring on ISFET response constitute the focal theme of discussion.

toward the efforts made in modifying ISFET gates with nanoparticle-based biomolecular preparations for biosensing applications. The use of carbon nanotubes for interfacing neurons with planar ISFETs is discussed in Section 10. The chapter concludes with summarizing remarks.

2. STRUCTURAL CONFIGURATIONS OF THE NANOSCALE ISFET

2.1. THE NANOPOROUS SILICON ISFET

Zehfroosh et al. (2010) reported a high-sensitivity ISFET in which nanoporous polysilicon was used on the thermally oxidized gate region (Figure 8.2b); the planar ISFET is shown in Figure 8.2a. This 1.5-μm-thick nanoporous poly-Si layer was formed by a sequential reactive-ion etching process consisting of two subcycles, in which the first passivation subcycle used a mixture of H_2/O_2 gases with a trace of SF_6 while the second etching subcycle used only SF_6. The three-dimensional nanoporous structures on the gate increase the adsorption surface area on the channel, which in turn couples the ion effects to the silicon dioxide layer below and then from the oxide layer to the silicon. The response of the nanoporous silicon ISFET to pH is not restricted to threshold voltage shifts. The slope of the drain current–reference voltage characteristics increases dramatically

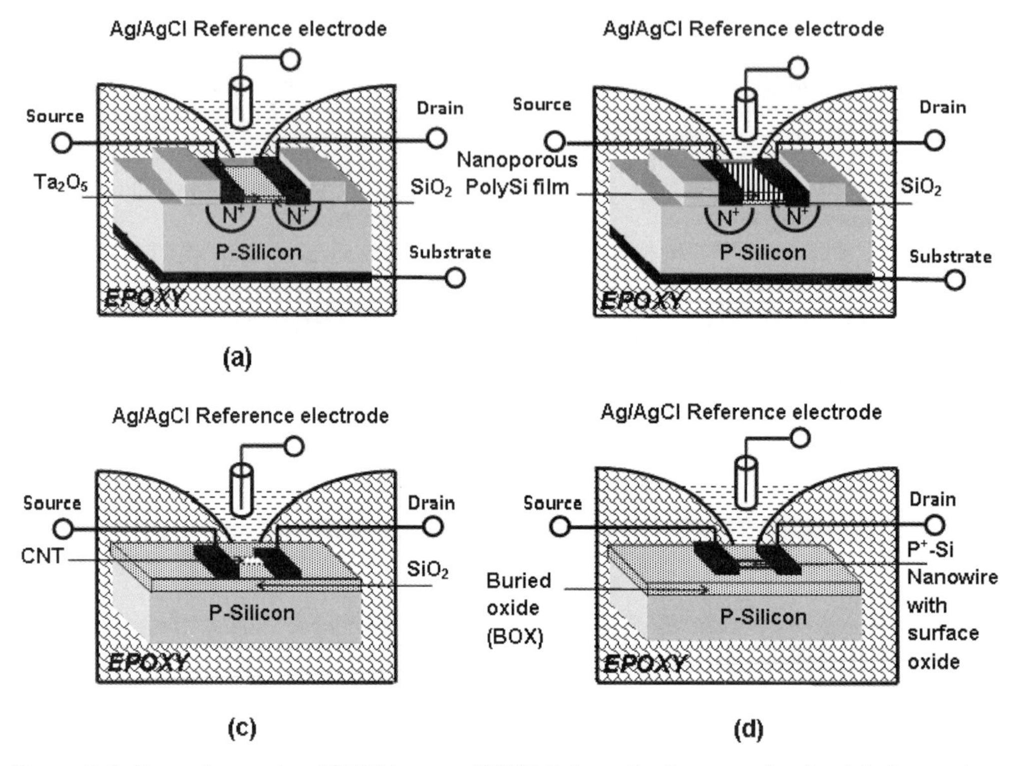

Figure 8.2. From planar micro-ISFET to nano-ISFET: Schematic diagrams showing (a) planar micro-ISFET; (b) nanoporous polysilicon ISFET; (c) CNT-ISFET; and (d) silicon nanowire ISFET. (Ideas from Zehfroosh et al. 2010; Dong et al. 2009, 2012; and Kim et al. 2011.)

with pH, giving high values of sensitivity. A high sensitivity, ~300 mV/pH, was reported, and ascribed to the 3-D nanostructures.

2.2. THE CNT ISFET

Carbon nanotubes (CNTs) have well-proven electrical, mechanical, and thermal properties. Dong et al. (2012) fabricated ISFETs by an atomic force microscopy (AFM)–based nano scratching process and using dielectrophoresis (DEP) to align the CNTs between the source and the drain to function as a nanochannel (Figure 8.2c) . By DEP, the CNT is translated in a suspending medium by applying an electric field. The electrical properties of CNTs are determined by current-sensing AFM (CSAFM). The *I*–*V* characteristics of CNTs were tested to verify their metallic or semiconducting character.

First the FET structure was fabricated and then CNTs were aligned by DEP. The basic FET structure consisted of four layers, from the bottom up, a Si wafer,

30 nm SiO_2, 20 nm Cr, and 50 nm Au, fabricated by micro-electro-mechanical systems (MEMS) surface micromachining techniques.

2.3. THE Si-NW ISFET

Kim et al. (2011) described the fabrication of a Si-NW ISFET (Figure 8.2d) by electron-beam lithography and semiconductor processing technology on silicon-on-insulator wafers consisting of a 100-nm-thick top boron-doped silicon layer (10^{15} cm^{-3}) with a 200-nm-thick buried oxide layer and a relatively thick substrate. The nanowire was 50 nm wide and 10 μm long. Arsenic implantation was done to form ohmic source and drain contacts, followed by Ti-Ag metallization.

The relative merits and demerits of the CNT ISFET vis-à-vis the Si-NW sensor are shown in Table 8.1.

Table 8.1. CNT and Si-NW ISFETs

Sl. No.	CNT ISFET	Si-NW ISFET
1.	Uses a SWCNT as the channel	Uses a silicon nanowire as the channel.
2.	Difficulty in selecting the semiconducting SWCNT for ISFET.	Doping is a routine semiconductor process.
3.	Manipulation and alignment of CNT between source and drain is done with atomic force microscopy, and the process is not amenable to bulk manufacturing.	Backed up by the vast silicon technological expertise, offering precise large-scale production at low cost.

3. PHYSICS OF THE Si-NW BIOSENSOR

3.1. BASIC PRINCIPLE

In comparison to micro-ISFETs, the Si-NW ISFET is simple in operation because it does not require the usual channel formation (Fan and Lu 2005, 2006; Chiesa et al. 2012). The N-channel micro-ISFET has an in-built N^+PN^+ structure to separate bulk silicon conduction from channel conduction, but this is not necessary in the Si-NW ISFET. The metallic films form ohmic contacts with the Si-NW.

Let us consider a conventional ISFET with channel length L μm and channel width W μm. When immersed in an analyte, it passes a certain drain–source current I_{DS}. This drain–source current represents the averaged effect of a large number of analyte molecules because the channel is very wide. If now the analyte concentration is decreased by a miniscule value of a few molecules per liter, the drain–source current changes accordingly by an amount ΔI_{DS}. At a given analyte concentration, the change in drain–source current (ΔI_{DS}) will be insignificantly small as compared to I_{DS}, implying that the conventional microchannel ISFET has a background drain–source current (in the absence of analyte) below which it will respond feebly to analyte concentration changes. If I_{DS} is in μA and ΔI_{DS} is in nA or pA, then ΔI_{DS} will not be distinguishable from I_{DS}. To bring ΔI_{DS} and I_{DS} into the same range, let us shrink the ISFET channel width to nanometric range. In other words, in order that a smaller number of molecules will be able to evoke a response, let us reduce the background drain–source current value, fixed by the ISFET geometry and dimensions, to nanoscale. Then, the cross-sectional area of the channel will decrease, and the starting drain–source current will decrease in proportion, so that small I_{DS} variations will not be negligible in comparison to the initial I_{DS} value and will be easily detected. The relative magnitudes of the cross-sectional area responsible for current flow I_{DS} and that for current change ΔI_{DS} will become comparable, and therefore they will be easily identifiable. Thus the difficulty of a larger I_{DS} value of the microchannel ISFET overshadowing the I_{DS} changes at low concentrations of analyte has been avoided.

In addition, the planar microchannel ISFET is subject to the influence of the analyte charges only from its top surface, but the nanowire is also affected from its lateral sides, which makes it more effective in sensing. For a nanowire, a surface charge causes a change in surface potential which induces a change in the space charge, either through accumulation or depletion, in a significant portion of the cross-sectional area of the nanowire, leading to a measurable change in the nanowire current in comparison to its original current.

3.2. ANALOGY WITH THE NANOCANTILEVER

An analogy will help in appreciating how this dimensional shrinkage will influence ISFET performance. The situation can be compared to cantilevers, which are excellent mass-sensing devices (Figure 8.3). The smaller the mass of the cantilever, the lower will be the additional detectable mass (Sone et al. 2004). If the cantilever itself is heavy, smaller masses will have imperceptible effect on the overall mass.

The mass change (Δm) is calculated from the resonance frequency shift ($\Delta f = f_2 - f_1$) by the equation (Sone et al. 2004)

$$\Delta m = -2m\left(\frac{\Delta f}{f}\right) \tag{8.1}$$

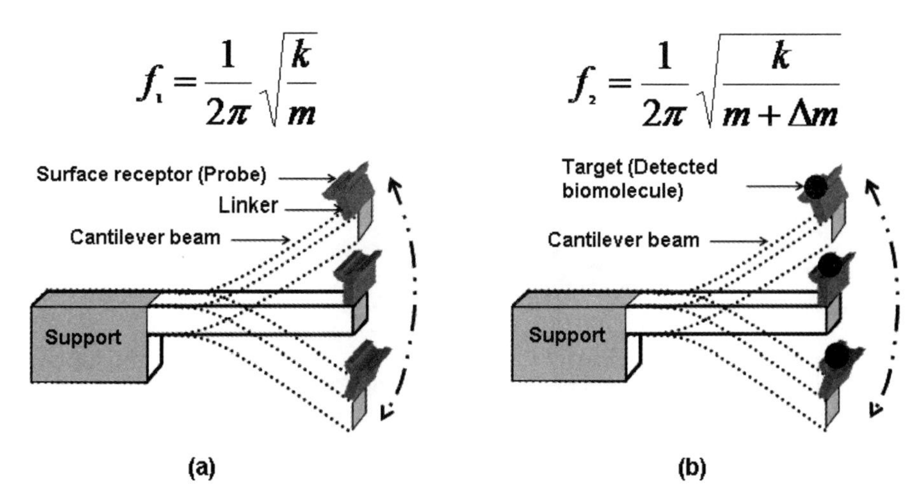

Figure 8.3. Diagrams illustrating the measurement of a small mass from the resonance frequency shift of a nanocantilever beam: (a) cantilever oscillation with a receptor fixed on the beam; and (b) cantilever oscillation after attachment of target biomolecule to the receptor. k is the spring constant of the cantilever, m is its initial effective mass, and $(m + \Delta m)$ is the final effective mass. The initial mass m must be small to enable detection and measurement of a small Δm. When the additional mass is comparable to the beam mass, the beam shows perceptible frequency shift; otherwise the frequency remains unaffected.

where the negative sign indicates that the resonance frequency decreases from f_1 to f_2, as the effective mass of the cantilever increases from m to $m + \Delta m$. The equation suggests that for detecting a small mass, the original effective mass m of the cantilever must also be small. Further, the cantilever should have a high resonance frequency. In conjunction with the above, an instrument is required which can measure a small frequency change with high resolution.

Similarly, in an ISFET, a nano-dimension channel is better than a micro-channel for sensing low concentrations of analytes.

3.3. PRELIMINARY ANALYSIS OF MICRO-ISFET DOWNSCALING TO NANO-ISFET

To understand the effect of dimensional shrinkage on device performance, let us go through a simple derivation as follows.

The sensitivity of the device is defined as the ratio of the change in device conductance $|\Delta G| = |G - G_0|$ to the original device conductance G_0, that is,

$$S = \frac{|\Delta G|}{G_0} = \frac{|G - G_0|}{G_0} \tag{8.2}$$

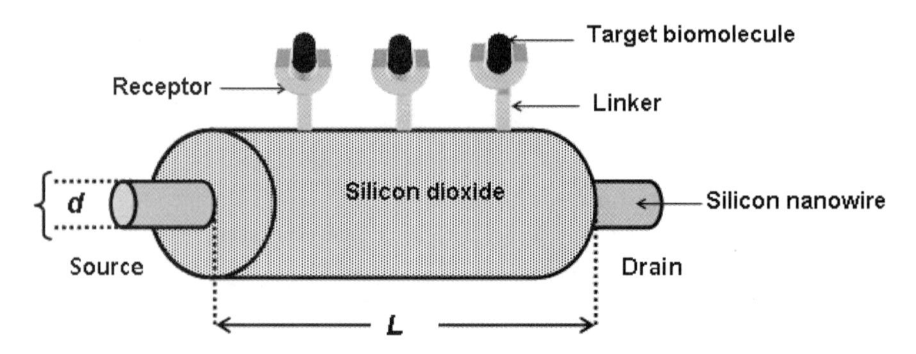

Figure 8.4. Idealized cylindrical nanowire showing different structural parameters. This wire consists of a silicon core cladded by thermal silicon dioxide all around the circumference. Practically realized nanowires differ from this ideal shape, which is assumed for simplifying calculations.

The nanowire is considered as a cylinder of diameter d, cross-sectional area A, and length L (Figure 8.4). Let the resistivity of nanowire material be ρ. The nanowire is made of p-type silicon with an acceptor concentration of N_A. Assuming complete ionization of impurities, the free carrier or hole concentration in the nanowire is N_A.

The initial conductance G_0 of the nanowire in the absence of surface charges is

$$G_0 = \frac{1}{R_0} = \frac{1}{(\rho L/A)} = \frac{1}{(1/N_A q\mu)\times\left[L/\left(\pi d^2/4\right)\right]} = \frac{N_A q\mu\pi d^2}{4L} \tag{8.3}$$

where R_0 is the initial resistance of the nanowire, q is the elementary charge, and μ is the carrier mobility in the nanowire. Also, the relation

$$\text{Resistivity } \rho = 1/(\text{carrier concentration } N_A \times \text{elementary charge } q \times \text{mobility } \mu) \tag{8.4}$$

has been applied.

Now, suppose the nanowire is immersed in the analyte solution. If N_S is the surface density (number of bound charged biomolecules/area of the surface) on the nanowire, the total charge causing accumulation or depletion of carriers in the nanowire below is given by

$$\Delta Q = qN_S \times \pi dL \tag{8.5}$$

The end surfaces of the nanowire with surface area $\pi d^2/2$ are not included because surface charges do not deposit there. Moreover, it must be pointed out that the oxide layer on the nanowire makes ionic interactions with the aqueous solution dependent on the pH of the solution. This will generate an additional charge density on the surface. However, if the experiments are always performed in an analye solution of constant pH or ionic concentration, the associated surface charge density with these ionic exchanges at the oxide/solution interface will remain constant and the biomolecular concentration alone will be responsible for nanowire conductance changes. Nonetheless, these ionic interactions are not inconsequential and therefore they will be dealt with separately.

The change in carrier concentration in the nanowire caused by this surface charge density is

$$\Delta n = (\Delta Q/q)/\text{volume of nanowire}$$

$$= \frac{\left[(qN_S \times \pi dL)/q\right]}{\left(\pi d^2/4\right)L} = \frac{4N_S}{d} \quad (8.6)$$

The final conductance of the nanowire with the attached biomolecules is obtained by simply adding Δn to N_A in the equation for G_0; hence

$$G = \frac{(N_A + \Delta n)\,q\mu\pi d^2}{4L} \quad (8.7)$$

From Eqs. (8.3) and (8.7), the change in nanowire conductance is found as

$$\Delta G = G - G_0 = \frac{(N_A + \Delta n)\,q\mu\pi d^2}{4L} - \frac{N_A q\mu\pi d^2}{4L} = \frac{\Delta n q\mu\pi d^2}{4L} \quad (8.8)$$

Substituting for Δn from Eq. (8.6), we get

$$\Delta G = G - G_0 = \frac{(4N_S/d)\times q\mu\pi d^2}{4L} = \frac{\pi q\mu N_S d}{L} \quad (8.9)$$

Hence, the sensitivity is

$$S = \frac{|\Delta G|}{G_0} = \frac{\pi q\mu N_S d/L}{N_A q\mu\pi d^2/4L} = \left(\frac{\pi q\mu N_S d}{L}\right)\times\left(\frac{4L}{N_A q\mu\pi d^2}\right) = \frac{4N_S}{N_A d} \quad (8.10)$$

For a given doping density of nanowire and biomolecular surface charge density, N_A and N_S are constants. Thus sensitivity scales inversely with nanowire diameter, meaning that nanowire diameter must be minimized to maximize sensitivity.

Table 8.2 presents a comparative view of planar and Si-NW ISFETs.

Table 8.2. Planar and Si-NW ISFETs

Sl. No.	Planar ISFET	Si-NW ISFET
1.	Micro-device	Nano-device
2.	Applies a gate voltage to create a channel.	The channel is already available in the form of the nanowire, i.e., the whole nanowire is a conductive channel.
3.	Gating effect is limited to top surface of the device.	Gating is done from the top surface of the nanowire; option for gating from its two sides is also available.
4.	Single-gate structure.	Multigate structure helps in tuning the device operation according to needs.
5.	The unknown molecules are attached to the receptors fixed on top of the gate plane, from only one side.	The unknown molecules are captured by receptor sites on the gate from all sides except the base.
6.	Lower surface-to-volume ratio of microchannel.	Higher surface-to-volume ratio of nanochannel.
7.	Able to detect larger concentrations of charges, in mM and nM ranges	Detection capability for lower concentration, in pM and fM ranges.
8.	Slower in response due to higher capacitances associated with larger features.	Faster response time because size reduction decreases the electrostatic capacitances.
9.	Higher power consumption.	Lower power consumption.

3.4. SINGLE-GATE AND DUAL-GATE NANOWIRE SENSORS

The preceding discussion does not mention the operational mode of the NW sensor. The nanowire core with its silicon dioxide mantle has been exposed to the analyte, and the variations in nanowire current have been observed with analyte concentration changes. Thus, unlike a planar ISFET, a nanowire may be used without the creation of a channel or inversion layer. The nanowire itself constitutes the nanochannel.

For more operational flexibility, a double-gate nanowire sensor has been devised (Stern 2007). Its schematic diagram is shown in Figure 8.5b along with single-gate nanowire sensor in Figure 8.5a. The two gates are called by various names, such as upper gate and lower gate, front gate and back gate, or liquid gate and back gate. The back gate connection taken from the substrate side is accomplished by a metallic wire to the power supply, whereas the liquid gate is fed through a platinum wire dipped in the solution. A calomel reference electrode is

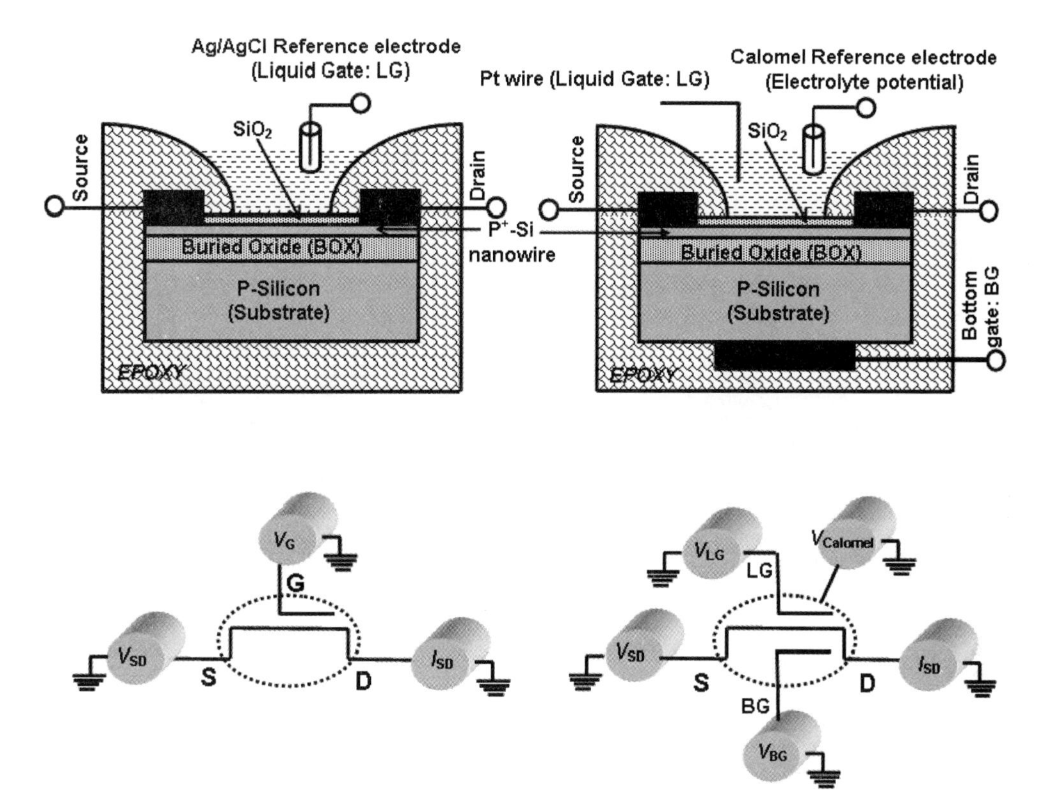

Figure 8.5. Cross-sectional diagrams with circuit symbols of two common silicon nanowire ISFET configurations: (a) single-gate nanowire sensor; (b) dual-gate nanowire sensor. (Idea from Knopfmacher 2011.)

used for measuring the electrolyte potential. Contact pads, bonding wires, and all electrical connections immersed in the solution are protected by insulating epoxy.

3.5. ENERGY-BAND MODEL OF THE NW SENSOR

Let us focus our attention on a *p*-type silicon NW FET, with an applied drain–source voltage V_{DS}, resulting in a drain–source current I_{DS}. An appropriate voltage V_{BG} is also applied to the substrate or back gate contact. The back gate allows the tuning of the device sensitivity, as we shall see later in Section 7.

The energy-band diagrams of the *p*-type Si NW (Carlen and van den Berg 2007) are displayed in Figure 8.6 in four conditions: (1) in the initial condition, in the absence of any charges on the SiO_2 surface; (2) in the second condition, in which a positive charge density is placed on the SiO_2 surface; (3) in the third condition, in which a small negative charge density is located over the SiO_2 surface;

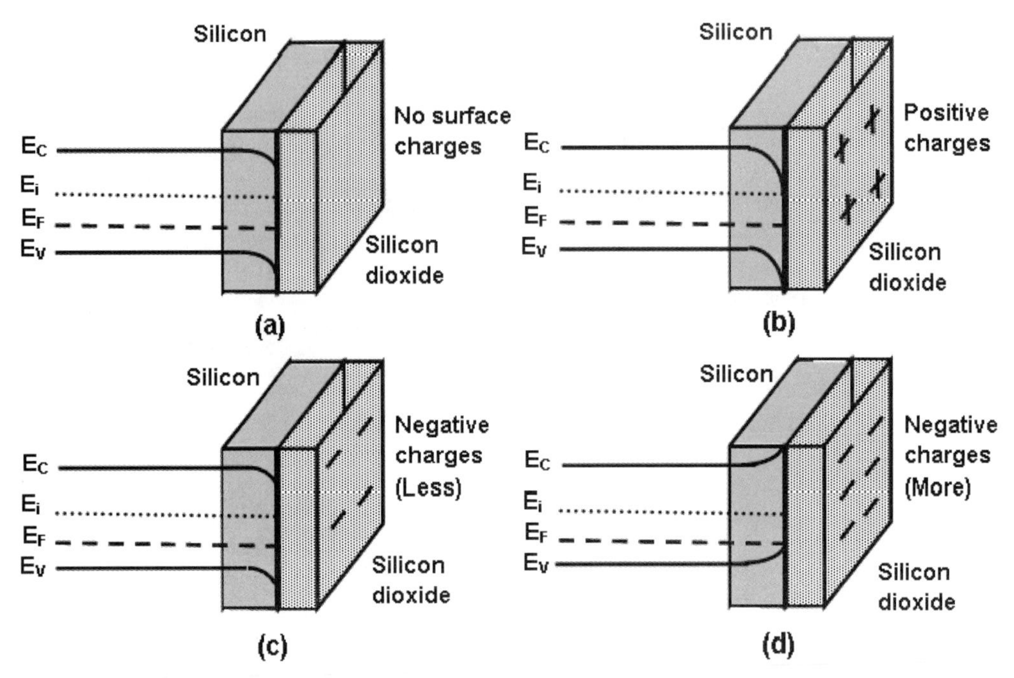

Figure 8.6. Energy-band diagrams of silicon–silicon dioxide structure under the influence of different surface charge densities: (a) Without any surface charge; the bulk and interface oxide charges are responsible for the bending of the energy bands in the downward direction. (b) Carrying a positive surface charge. (c) Carrying a small negative surface charge, insufficient to overcome the effect of oxide charges. (d) At higher negative surface charge density exceeding the effect of oxide charges. (Idea from Carlen and van den Beg 2007.)

and (4) in the fourth condition, in which a larger negative charge density is situated over the SiO_2 surface .

Prima facie in the first condition of zero charge on the SiO_2 surface, it appears that the band lines will be straight, which never happens in practice. One must not forget that trapped positive charges reside in the silicon dioxide. Also, interface charges, generally positive, are present at the oxide/silicon interface. The combined effect of these positive charges in the bulk SiO_2 and at the interface is repulsion of majority carrier holes in *p*-type Si NWs away from the silicon surface and attraction of minority carrier electrons toward the silicon surface. Consequently, the holes are depleted near the surface. Usually, the oxide charges are not large enough to attract sufficient electrons to invert the carrier population at the surface. Thus, the main effect of oxide charges is the formation of a depletion region beneath the oxide surface, which is shown in the band diagram in Figure 8.6a by a downward bending of the bands.

The second condition, placing a positive charge density on the silicon dioxide surface, further increases the downward band bending produced in the first condition. Thus a proper band-diagram representation of this case is provided by a relatively larger bending of bands in the downward direction in Figure 8.6b, as compared to that in Figure 8.6a.

In the third condition, a negative charge density on the oxide surface, the holes are pulled upward by electrostatic attraction toward the silicon surface by negative surface charges, while the electrons are pushed away from the silicon surface in the downward direction. This leads to the creation of an *accumulation region* at the Si surface in which the hole density is higher than in the NW portion below. An increase in the hole population under the oxide layer is shown in the band diagram by an upward band bending, in contrast to the downward band bending in the preceding two conditions. This upward band bending nullifies the previous downward band bending to some degree, so that the net result is shift of the energy bands upward near the silicon surface but still the bands may be bent slightly downwards, as the downward band bending is not fully annulled (Figure 8.6c). The upward band bending depends on the extent to which the negative charge density has been successful in compensating the effect of positive charges.

In the fourth condition, when the effect of negative charges overcomes and exceeds that of the positive charges, the bands may ultimately bend upward, as shown in the energy-band diagram in Figure 8.6d.

4. NAIR-ALAM MODEL OF Si-NW BIOSENSORS

4.1. THE THREE REGIONS IN THE BIOSENSOR

Nair and Alam (2007, 2008) partitioned the system into three regions as follows (Figure 8.7).

1. **The silicon nanowire:** A cylindrical nanowire geometry with a diameter d is assumed. The carrier transference in the nanowire takes place according to the well-known drift-diffusion equations. The potential ϕ and carrier concentrations in the nanowire are interrelated by the Poisson equation.
2. **The silicon dioxide film:** A perfect oxide layer is assumed, without any bulk oxide or silicon–silicon dioxide interface traps. As before, the potential in the oxide is given by the Poisson equation.
3. **The fluidic environment:** This consists of the electrolyte containing the ions and the biomolecules, and serves as the buffer medium for target–receptor binding. In the presence of the electrolyte, the potential ϕ in the fluid is obtained from the Poisson-Boltzmann equation.

As already pointed out in Section 3.3, a surface charge density is produced by protonation of silanol groups at the SiO_2/water interface. This charge density

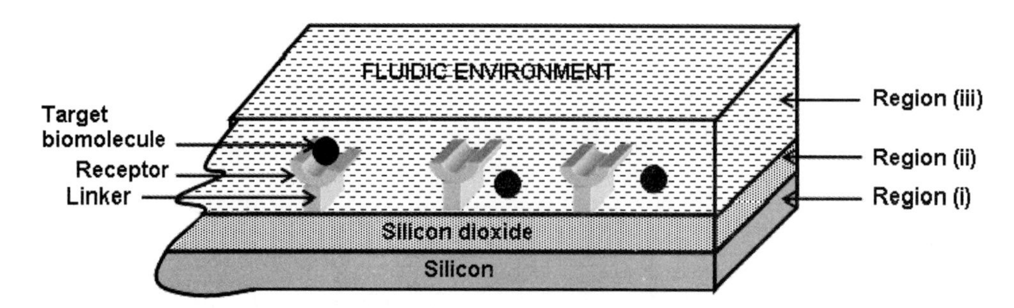

Figure 8.7. Illustrating the three regions of interest for modeling silicon nanowire biosensors, namely, the oxide layer covering the nanowire, the silicon nanowire itself, and the aqueous electrolyte solution containing the target biomolecules. The receptors are shown anchored over the silicon dioxide surface for capturing the charged target molecules, whereby modulation of conductance of the silicon nanowire produces a measurable detection signal. (Idea from Nair and Alam 2007.)

remains constant if the ionic concentration remains the same during the detection. In a first-order analysis, this surface charge density is assumed to be zero. Its effect will be studied later.

4.2. COMPUTATIONAL APPROACH

The model takes Eqs. (8.11)–(8.14) as the premises and solves them numerically under two conditions (Figure 8.8).

Equations (8.11) and (8.12) are the 3-D drift-diffusion equations used for describing carrier transport in semiconductor physics:

$$\mathbf{J}_n = qn\mu_n\mathbf{E} + qD_n\nabla n \tag{8.11}$$

$$\mathbf{J}_p = qp\mu_p\mathbf{E} + qD_p\nabla p \tag{8.12}$$

In these equations, $\mathbf{J}_n$ and $\mathbf{J}_p$ are the electron and hole currents, respectively, under an applied electric field of intensity $\mathbf{E}$; q is the electronic charge = 1.6×10^{-19} C; n and p are electron and hole concentrations; μ_n, μ_p are electron and hole mobilities; and D_n, D_p are electron and hole diffusion coefficients.

Equation (8.13) is the Poisson equation relating the electrostatic potential $\phi(r)$ at the point r to the sum-total charge,

$$\nabla.\nabla\varphi(r) = -\left(\frac{q}{\varepsilon_0\varepsilon_{\text{Si}}}\right)(p - n + N_D - N_A) \tag{8.13}$$

where ε_0 is free-space permittivity, ε_{Si} is the dielectric constant of silicon, and N_D, N_A are the concentrations of donor and acceptor impurities, respectively.

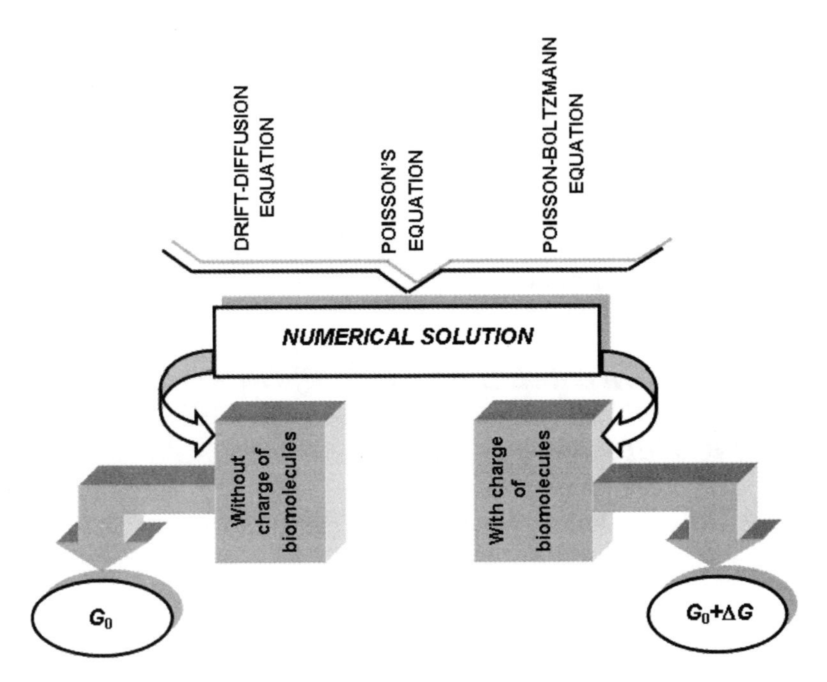

Figure 8.8. Simulation methodology for conductance variation of Si-NW sensor with biomolecule capturing.

Equation (8.14) is the Poisson-Boltzmann equation (PBE), the fundamental equation arising in the Debye-Hückel theory (Zhou and Wu 2012), derived from a continuum model of the solvent and counter-ion environment surrounding a biomolecule:

$$\nabla\left[\varepsilon_w \nabla\phi(r)\right] = \kappa^2 \sinh(\beta\phi) - q\sum_{i=1}^{N} z_i \delta(r - r_i) \tag{8.14}$$

where ε_w is the dielectric constant of the electrolyte; κ represents the modified Debye-Hückel parameter, which includes the ionic strength of the solution and the accessibility of the ions to the interior of the solute, where

$$\kappa^2 = \frac{2q^2 I_0 N_{\text{Avogadro}}}{kT} \tag{8.15}$$

in which I_0 is the ionic concentration in molar units, N_{Avogadro} is Avogado's number, the hyperbolic sine term stands for the contribution of a 1–1 electrolyte, e.g., Na^+–Cl^-, in which the ions obey the Boltzmann distribution. The symbol N is the number of point charges, r_i refers to their positions, z_i to magnitudes of the charges, and $\delta(r)$ is the Dirac delta function (Heitzinger and Klimeck 2007; Wang 2010; Wang and Li 2010).

In the initial or starting condition, the charge of the attached biomolecules on the right-hand side of Eq. (8.14) is set to zero. This gives the conductance G_0 of the nanowire before capturing the biomolecule. In the final condition, the charge of the attached biomolecule is taken into consideration by approximating the charge of the biomolecule as a surface charge density which extends over a finite distance. This yields the nanowire conductance G after biomolecule capture. The difference between the final and the initial conductances represents the conductance change $|\Delta G| = G - G_0$, which divided by G_0 gives the sensitivity S of the sensor.

4.3. EFFECT OF NANOWIRE DIAMETER (d) ON SENSITIVITY AT DIFFERENT DOPING DENSITIES, WITH AIR AS THE SURROUNDING MEDIUM

Nair and Alam (2007) simulated the sensitivity parameter S as a function of nanometer diameter from 10 to 40 nm at dopant concentrations of 10^{17}, 10^{18}, and 10^{19} cm^{-3}. For these simulations, the nanowire was placed in air so that the surrounding medium was air. They found that the highest sensitivity values were achieved at 10^{17} cm^{-3}, and the sensitivity decreased by about three orders of magnitude at 10^{19} cm^{-3}. Further, the highest sensitivity was for nanowires of diameter 10 nm, and it declined with increasing diameter of the nanowire.

4.4. EFFECT OF NANOWIRE LENGTH (L) ON SENSITIVITY AT DIFFERENT DOPING DENSITIES, WITH AIR AS THE SURROUNDING MEDIUM

As before, the sensitivity of the nanowire was calculated for different lengths from $L = 0.25$ to 2 μm at the above dopant concentrations considering air as the surrounding medium. The interesting observation was that the length of the nanowire also affected the sensitivity: The greater the length, the less was the sensitivity.

The sensitivity equation can be rewritten in terms of initial resistance R_0 and the change in resistance ΔR as $S \sim \Delta R/R_0$. Although ΔR remains the same regardless of the length of the nanowire, R_0 and hence S scale with L. However, doping density had the same effect on sensitivity, viz., a lower doping density resulting in a higher sensitivity value.

4.5. EFFECT OF THE FLUIDIC ENVIRONMENT

The fluid surrounding the biosensor is not air but water, which has a high dielectric constant (= 78.54 at room temperature). Therefore, the effect of water must be incorporated to simulate a realistic situation. The presence of water influences the induced charge profile defined by the shape and extent of the depletion/accumulation region produced by the charged biomolecules.

4.5.1. Effect of NW Operational Modes (Depletion/Accumulation) on Sensitivity, with Air or Water as Surrounding Medium

When a localized positive charge Q^+ is placed on the *p*-type NW surface, the high dielectric constant of water causes spreading of the fringing field lines. As the NW surface goes into depletion, the fringing field lines in water allow the depletion layer to stretch more toward the sides, so that the depletion region is shallower in depth but longer in length. As opposed to this, the fringing field spreading is less in air; hence the depletion region in air is restricted to a smaller length but extends over a greater depth. With the NW in depletion, ΔR for a NW in water will be less than ΔR in air. This happens because a charge Q^+ causes a resistance change $\Delta R \sim Q^+/[\pi d^2(d^2 - 4W_d^2)]$, where W_d is the width of the depletion region, which changes upon placing Q^+. The ΔR expression shows that ΔR is affected by the effective diameter changes, not by length variation, and diameter changes are more pronounced when the nanowire is surrounded by air, where the depletion region penetrates deeper into the nanowire. Since $S \propto \Delta R$, we can write

$$S_{\text{water}} \leq S_{\text{air}} \qquad \text{for NWs in depletion} \tag{8.16}$$

For a negative charge Q^- on the NW surface, there will be accumulation of majority carrier holes near the NW surface and this accumulation layer extends up to a depth δ = Debye length $(\lambda_D)_{\text{Si}}$. ΔR will now depend mainly on accumulation length; the depth being the Debye length, effective diameter changes do not occur. The reason is that a charge Q^- causes a resistance change $\Delta R \sim Q^-/\pi d^2(d^2 + \xi L_a^{-2})$ where ξ is a constant and L_a is the accumulation length. Due to the greater spread of fringing field lines in water than in air, the accumulation will take place over a larger length in water than in air, so ΔR for NWs in water will be greater than ΔR in air. From, $S \propto \Delta R$, we have

$$S_{\text{water}} \geq S_{\text{air}} \qquad \text{for NWs in accumulation} \tag{8.17}$$

Thus for depletion-mode operation, sensitivity in air is greater than sensitivity in water, whereas for accumulation-mode operation, sensitivity in water is greater than sensitivity in air. Hence for biomolecular charge detection in air, such as for chemical gas sensors, the NW doping polarity should be chosen such that the sensor works in depletion. Contrarily, for biomolecular charge detection in water, e.g., for biosensors, the NW doping polarity should be selected in such a way that the sensor works in accumulation. However, in the case of water, the effect of doping polarity is comparatively less pronounced than in air, where an order-of-magnitude ratio between sensitivity in depletion and accumulation modes is noticed. Instead, for water, dopant concentration is found to be a more critical parameter affecting the sensitivity than operation in depletion or accumulation mode. Therefore, it is the doping density, not the mode, which controls sensitivity.

4.5.2. Effect of Electrolyte Concentration on NW Detection Capability

Screening of the charges of target biomolecules by ions lowers the effective charge modulating the NW conductance. Hence, the finite size of biomolecules, their distance of separation from the NW, and the ionic solutions used for target–receptor binding all play important roles in NW behavior.

The experiments of Stern et al. (2007) on molecular charge screening by counter-ions present in the analyte solution emphasize the significance of accounting for the screening effect in designing protocols for Si NW sensors. Undeniably, all aqueous solutions, even water in pure form, contain ions, which terminate electric field lines, and therefore severely screen Coulomb interactions among charges. To elaborate how this screening effect interferes with the measurements, suppose a positively charged ion species is being investigated. Electrostatic interactions will compel the negative ions present in the supporting electrolytic solution (in which the analyte molecules are suspended) to come into the vicinity of the positive ion species, surrounding the latter. This surrounding of positive species by the negative ion cloud in solution will take place on a length scale called the *Debye screening length* $(\lambda_D)_{\text{Electrolyte}}$ such that the electrostatic potential produced by the positive species will decay with distance to zero rapidly according to the exponential function (Table 8.3). Thus free ions not only reduce the overall amplitude of Coulomb interactions, they also change the shape of the potential energy gradient, making it fall to zero exponentially beyond a characteristic distance called the *Debye screening length.*

The Debye screening length depends on the sum-total charges of ions present in the solution. For solutions in water at room temperature, it is convenient to express the inverse Debye length in terms of the Bjerrum length $l_B = 0.7$ nm,

$$\lambda_D^{-1} = \sqrt{4\pi l_B \sum_i \rho_i z_i^2} \tag{8.18}$$

where ρ_i is the density and z_i is the valence of the ions. The screening length provides a realistic first estimate of the distance beyond which Coulomb interactions can be ignored. It also gives an idea about the size of the region near a point charge where oppositely charged ions can be found. It is easily seen that as more

Table 8.3. Debye screening lengths in electrolyte and semiconductor

FEATURE	DEBYE LENGTH IN AN ELECTROLYTE $(\lambda_D)_{Electrolyte}$		DEBYE LENGTH IN A SEMICONDUCTOR $(\lambda_D)_{Si}$	
Definition	A characteristic length of the electrolytic system, giving the distance at which the charge density and the electrical potential of an ion atmosphere fall to e^{-1}.		A characteristic length for electrostatic problems in semiconductors, giving the distance over which local electric field affects distribution of free charge carriers.	
Equation	$(\lambda_D)_{Electrolyte}^{-1} = \sqrt{\frac{\varepsilon_0 \varepsilon_w kT}{2N_{Avogadro} q^2 I}} = \frac{9.66}{\sqrt{I}} \text{nm}$ where ε_0 is the permittivity of free space, ε_w is the dielectric constant of water, k is the Boltzmann constant, T is the absolute temperature, $N_{Avogadro}$ is the Avogadro number, q is the elementary charge, and I is the ionic strength of the electrolyte (mole/m^3).		$(\lambda_D)_{Si} = \sqrt{\frac{\varepsilon_0 \varepsilon_{Si} kT}{q^2 N}} = \frac{40.92 \times 10^{11}}{\sqrt{N}} \text{nm}$ where ε_{Si} is the dielectric constant of silicon, N is the dopant concentration (acceptor or donor in m^{-3}), and other symbols have their usual meanings.	
Typical values at room temperature	Ionic strength of 1–1 electrolyte (mole/m^3)	$(\lambda_D)_{Electrolyte}$ (nm)	Dopant concentration (m^{-3})	$(\lambda_D)_{Si}$ (nm)
	0.001	305.48	10^{22}	40.92
	0.01	96.60	10^{23}	12.94
	0.1	30.55	10^{24}	4.09
	1.0	9.66	10^{25}	1.29

ions are added to a solution, the screening length decreases because the valences appear in squared form (z_i^2) in the equation.

In their experiments, Stern et al. (2007) noticed that for a starting ionic strength of buffer solution, for which $(\lambda_D)_{\text{Electrolyte}}$ was ~7.3 nm, almost the whole charge present on the analyte remained unscreened, as evidenced by the high source–drain current. Upon increasing the ionic strength of the buffer solution by one order of magnitude, $(\lambda_D)_{\text{Electrolyte}}$ decreased to ~2.3 nm, resulting in partial screening of analyte charges. Further 10-fold increase in buffer ionic strength completely screened the charges of the analyte, as indicated by source–drain current falling to the baseline value. Upon decreasing the buffer ionic strength, reversal of ionic screening and progressive restoration of current to previous higher values was observed. Thus they convincingly proved that choice of a suitable Debye length $(\lambda_D)_{\text{Electrolyte}}$ value by properly controlling the ionic strength of buffer solution was essential to carry out a logical sensing experiment. At millimolar ionic concentrations, about half the biomolecular charge is screened by the ions, so detection strategies must be planned cautiously to avoid higher concentrations of ions.

Nair and Alam (2006, 2008) self-consistently solved diffusion-capture equations and the Poisson-Boltzmann equations to determine the screening-limited response of nanobiosensors. Diffusion of target molecules present in the electrolyte toward the receptors is described by the diffusion equation

$$\frac{d\rho}{dt} = D\nabla^2\rho \tag{8.19}$$

where ρ is the concentration of the target molecules and D is their diffusion coefficient. For the capturing of molecules by the receptors, the capture equation is

$$\frac{dN}{dt} = k_F\left(N_0 - N\right)\rho_S - k_R N \tag{8.20}$$

where N_0, N are the initial number of receptors and the number of conjugated receptors; k_F, k_R are the forward and reverse reaction constants, viz., capture and dissociation constants; and ρ_S is the surface concentration of analyte molecules at an instant of time t. Because of less probability of deconjugation in biosensing, the above equation is approximated to

$$\frac{dN}{dt} \approx k_F N_0 \rho_S \tag{8.21}$$

Nair and Alam (2008) combined analytical solutions of diffusion-capture and the Poisson-Boltzmann equations, and showed that electrostatic screening by the ionic environment in the electrolyte constrains the sensitivity to

$$S(t) \approx c_1 \left\{ \ln(\rho_0) - \frac{\ln(I_0)}{2} + \frac{\ln(t)}{D_F} + c_2 [\mathrm{pH}] \right\} + c_3 \quad (8.22)$$

where c_1, c_2, c_3 are constants dependent on the sensor geometry, ρ_0 is the equilibrium concentration of target molecules, I_0 is the ionic strength of the electrolyte, and D_F is the fractal dimension of the sensor surface. *Fractal dimension of a surface* is a statistical index indicating the extent of fractality of the surface. It quantifies how much the fractal surface fills the spatial volume; $D_F = 1$ for cylindrical nanowire and $D_F = 2$ for planar sensors.

Main predictions of the model are that: (1) the sensitivity increases logarithmically with concentration of the target molecules; (2) the sensitivity decreases logarithmically with ionic strength of the electrolyte; (3) the transient response of the sensor increases logarithmically with time; and (4) the sensitivity increases linearly with the pH of the solution.

In view of (3) above, Nair and Alam (2008) pointed out that the molecular screening effect increases the incubation time to reach the same value of sensor output. This makes it mandatory to formulate functionalization schemes at low ionic strengths to prevent screening from affecting the magnitude of the response and also to obtain a signal of sufficient strength in a reasonable time span.

Liu et al. (2008) studied the role of screening in Si-NW sensors by using deep-level device simulation capabilities. They showed that by introducing electrodiffusion flow in the electrolyte, the screening effects were significantly reduced so that the signal strength was increased by a factor ≥ 10. This was suggested as a method to overcome the nanowire performance limitation by screening.

Liu et al. (2012) proposed a multiphysics model to study the contribution of electrokinetic parameters such as electrophoretic force and electro-osmotic flow on the biomolecular detection mechanism. They established that in a single nanowire–based sensor, electrokinetic effects could decrease the detection time over 90 times, compared with that by pure biomolecular diffusion. Further reduction was possible by increasing the applied gate voltage or the number of nanowires, suggesting that that faster biomolecular detection at ultralow concentration was achievable by appropriate combinations of electrokinetic effects and nanowire sensor design.

4.6. OVERALL MODEL IMPLICATIONS

At first sight, it appears that the sensitivity of NW sensors can be indefinitely increased by reducing the doping density, wire diameter, and length, but at low dopant concentrations, random dopant fluctuations become extremely high, e.g., at a concentration of 10^{17} cm^{-3} in a 10-nm-diameter NW, there are only 8 dopant atoms per micrometer of sensor length, with variability in distribution, so that localized dopant concentration at the point of attachment of biomolecules is a decisive sensitivity-determining factor. Moreover, the nanowire diameter is reducible to typically 20–50 nm, below which $1/f$ noise increases, rendering this device incapable for sensing applications. The effects of water on sensitivity and ionic concentration of solution on detection capability should not be overlooked.

5. pH RESPONSE OF SILICON NANOWIRES IN TERMS OF THE SITE-BINDING AND GOUY-CHAPMAN-STERN MODELS

Thus far, the surface charge exchanges between silicon dioxide and electrolyte were ignored, with the postulate that experiments will be conducted under constant pH conditions. Chen et al. (2011) presented a comprehensive analytical model of a nano-ISFET based on the site-binding (SB) and Gouy-Chapman-Stern (GCS) models (Yates 1974). The SB model considers the surface reactions on the NW oxide dielectric to determine the surface response of the sensor, while the GCS model relates the surface response to the bulk electrolyte.

The treatment follows the footsteps of micro-ISFET theory. The silicon dioxide surface of the NW sensor undergoes hydration to form silanol (Si–OH) groups, which are one of three types, either positively charged, negatively charged, or neutral.

Deprotonation of Si–OH or protonation of $Si–O^-$ is described by the equation

$$\text{Si–OH} \leftrightarrow \text{Si–O}^- + \text{H}_S^+ \tag{8.23}$$

with an associated dimensionless dissociation constant

$$K_- = \frac{\nu_{\text{Si–O}^-} a_{\text{H}_S^+}}{\nu_{\text{Si–OH}}} \tag{8.24}$$

where ν's denote density of surface sites of the relevant species and $a_{\text{H}_S^+}$ stands for the surface activity or effective concentration of the hydrogen ion.

Deprotonation of $Si–OH_2^+$ is represented as

$$\text{Si–OH}_2^+ \leftrightarrow \text{Si–OH} + \text{H}_S^+ \tag{8.25}$$

with a dimensionless dissociation constant

$$K_+ = \frac{\nu_{\text{Si–OH}} a_{\text{H}_S^+}}{\nu_{\text{Si–OH}_2^+}} \tag{8.26}$$

The adsorbed hydrogen ions from the electrolyte create a surface charge density σ_0 on the oxide surface, given by

$$\sigma_0 = \text{difference between positive and negative ionic charge densities} \quad (8.27)$$

$$\begin{aligned}\sigma_0 &= q\nu_{Si\text{-}OH_2^+} - q\nu_{Si\text{-}O^-} \\ &= q\left\{\left(\frac{\nu_{Si\text{-}OH} a_{H_S^+}}{K_+}\right) - \left(\frac{K_- \nu_{Si\text{-}OH}}{a_{H_S^+}}\right)\right\} \\ &= \frac{q\left(\nu_{Si\text{-}OH} a_{H_S^+}^{\ 2} - K_- K_+ \nu_{Si\text{-}OH}\right)}{K_+ a_{H_S^+}} \\ &= \frac{q\nu_{Si\text{-}OH}\left(a_{H_S^+}^{\ 2} - K_- K_+\right)}{K_+ a_{H_S^+}}\end{aligned} \quad (8.28)$$

where q is the electronic charge.

Noting that the oxide surface can have any polarity, the number of surface sites N_s is written as

$$N_s = \text{number of neutral sites} + \text{number of positive ionic sites} + \text{number of negative ionic sites} \quad (8.29)$$

$$\begin{aligned}N_s &= \nu_{Si\text{-}OH} + \nu_{Si\text{-}OH_2^+} + \nu_{Si\text{-}O^-} \\ &= \nu_{Si\text{-}OH} + \frac{\nu_{Si\text{-}OH} a_{H_S^+}}{K_+} + \frac{K_- \nu_{Si\text{-}OH}}{a_{H_S^+}} \\ &= \nu_{Si\text{-}OH}\left(1 + \frac{a_{H_S^+}}{K_+} + \frac{K_-}{a_{H_S^+}}\right) \\ &= \nu_{Si\text{-}OH}\left(\frac{K_+ a_{H_S^+} + a_{H_S^+}^{\ 2} + K_- K_+}{K_+ a_{H_S^+}}\right)\end{aligned} \quad (8.30)$$

Dividing Eq. (8.28) for σ_0 by Eq. (8.30) for N_S, we get

$$\frac{\sigma_0}{N_S} = \frac{q\nu_{Si\text{-}OH}\left(a_{H_S^+}^{\ 2} - K_- K_+\right) / K_+ a_{H_S^+}}{\nu_{Si\text{-}OH}\left[\left(K_+ a_{H_S^+} + a_{H_S^+}^{\ 2} + K_- K_+\right) / K_+ a_{H_S^+}\right]} = \frac{q\left(a_{H_S^+}^{\ 2} - K_- K_+\right)}{K_+ a_{H_S^+} + a_{H_S^+}^{\ 2} + K_- K_+} \quad (8.31)$$

Hence,

$$\sigma_0 = \frac{qN_S\left(a_{H_S^+}{}^2 - K_-K_+\right)}{K_+a_{H_S^+} + a_{H_S^+}{}^2 + K_-K_+} \tag{8.32}$$

The effect of a small change in $a_{H_S^+}$ on σ_0 is given by

$$\frac{\partial\sigma_0}{\partial \mathrm{pH}_S} = \frac{\partial}{\partial \mathrm{pH}_S}\left[\frac{qN_S\left(a_{H_S^+}{}^2 - K_-K_+\right)}{K_+a_{H_S^+} + a_{H_S^+}{}^2 + K_-K_+}\right] = -q\beta_{\text{intrinsic}} \tag{8.33}$$

with

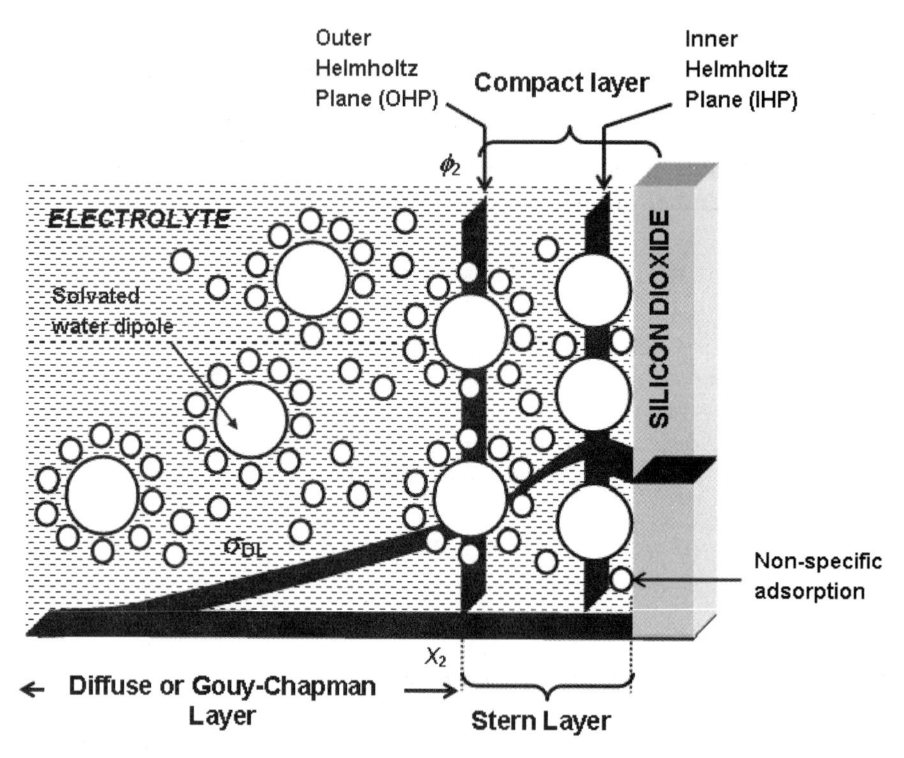

Figure 8.9. Schematics of the electrical double-layer formation at the silicon dioxide/electrolyte interface as a result of diverse phenomena taking place in this region. The Stern and diffuse layers as well as the inner and outer Helmholtz planes are shown.

$$\beta_{\text{intrinsic}} = -\frac{\partial}{\partial \text{pH}_S}\left[\frac{N_S\left(a_{H_S^+}{}^2 - K_-K_+\right)}{K_+a_{\text{H}_S^+} + a_{\text{H}_S^+}{}^2 + K_-K_+}\right] \tag{8.34}$$

$\beta_{\text{intrinsic}}$ is the *intrinsic buffer capacity* characterizing the ability of the surface to buffer small pH changes.

To counterbalance the above surface charge buildup, i.e., to ensure charge neutrality, an equal and opposite charge must exist in the electrolyte near the oxide surface. The electrical double layer (Attard 1996) set up at the oxide–electrolyte interface (Figure 8.9) comprises (1) the *surface charge* σ_0 with the accompanying potential ψ_0; (2) the *Stern layer* spanning the distance from the oxide surface to x_2, defining the plane of closest approach of ionic centers in the solution to the oxide surface; and (3) The *diffuse charge layer* σ_{DL} extending from x_2 to a distance far away from the oxide surface into the electrolyte.

The potential across the solution is given by

$$\psi_0 = \phi_2 + \sqrt{8kT\varepsilon_0\varepsilon_w n^{\circ}}\sinh\left(\frac{zq\phi_2}{2kT}\right)C_{\text{Stern}}{}^{-1} \tag{8.35}$$

where ϕ_2 is the potential at x_2, ε_0 is the permittivity of free space, ε_w is the dielectric constant of the solution, n° is the number concentration of each bulk ion, z is the ionic charge, and C_{Stern} is the Stern capacitance per unit area, written as

$$C_{\text{Stern}} \approx \frac{\varepsilon_0\varepsilon_w}{x_2} \tag{8.36}$$

The surface charge density σ_0 = (−) charge density in the double layer, i.e.,

$$\sigma_0 \approx -\sigma_{DL} = \sqrt{8kT\varepsilon_0\varepsilon_w n^{\circ}}\sinh\left(\frac{zq\phi_2}{2kT}\right) \tag{8.37}$$

The charge storage capability of the double layer in response to small potential changes is measured by the differential capacitance,

$$C_{\text{differential}} = \frac{\partial\sigma_0}{\partial\psi_0} \tag{8.38}$$

Combining the SNB and GCS models, the surface potential ψ_0 and the activity of bulk hydrogen ions $a_{\text{H}_B{}^+}$ are related to the activity of surface hydrogen ions $a_{\text{H}_S{}^+}$ through the Boltzmann distribution function as

$$a_{H_S^+} = a_{H_B^+} \exp\left(-\frac{q\psi_0}{kT}\right) \quad (8.39)$$

Since

$$pH_S = -\log_{10} a_{H_S}{}^+ = \log_{10}\left(\frac{1}{a_{H_S}{}^+}\right) = \frac{1}{2.303}\ln\left(\frac{1}{a_{H_S}{}^+}\right) \quad (8.40)$$

therefore,

$$\ln\left(\frac{1}{a_{H_S}{}^+}\right) = 2.303\,pH_S \quad (8.41)$$

or

$$a_{H_S}{}^+ = \exp(-2.303\,pH_S) \quad (8.42)$$

Hence,

$$\exp(-2.303\,pH_S) = \exp(-2.303\,pH_B) \times \exp\left(-\frac{q\psi_0}{kT}\right) \quad (8.43)$$

Taking the natural logarithm of both sides,

$$-2.303\,pH_S = -2.303\,pH_B - \frac{q\psi_0}{kT} \quad (8.44)$$

or

$$pH_B = pH_S - \frac{q\psi_0}{2.303kT} \quad (8.45)$$

Differentiation of Eq. (8.45) for pH_B with respect to pH_S gives

$$\frac{\partial pH_B}{\partial pH_S} = 1 - \frac{q}{2.303kT}\left(\frac{\partial\psi_0}{\partial pH_S}\right) = 1 + \frac{q\beta_{intrinsic}}{2.303kTC_{differential}} = \frac{2.303kTC_{differential} + q\beta_{intrinsic}}{2.303kTC_{differential}} \quad (8.46)$$

where the following substitution has been made:

$$\frac{\partial\psi_0}{\partial pH_S} = \frac{\partial\psi_0}{\partial\sigma_0} \times \frac{\partial\sigma_0}{\partial pH_S} = \left(\frac{1}{C_{differential}}\right) \times (-q\beta_{intrinsic}) = -\frac{q\beta_{intrinsic}}{C_{differential}} \quad (8.47)$$

Now, $\partial pH_S/\partial pH_B$ is

$$\frac{\partial pH_S}{\partial pH_B} = \frac{2.303kTC_{\text{differential}}}{2.303kTC_{\text{differential}} + q\beta_{\text{intrinsic}}} \tag{8.48}$$

The pH sensitivity of the surface in contact with the electrolyte is expressed as

$$\begin{aligned}
\frac{\partial \psi_0}{\partial pH_B} &= \frac{\partial \psi_0}{\partial pH_S} \times \frac{\partial pH_S}{\partial pH_B} = \frac{\partial \psi_0}{\partial \sigma_0} \times \frac{\partial \sigma_0}{\partial pH_S} \times \frac{\partial pH_S}{\partial pH_B} \\
&= \left(\frac{1}{C_{\text{differential}}}\right) \times (-q\beta_{\text{intrinsic}}) \times \frac{2.303kTC_{\text{differential}}}{2.303kTC_{\text{differential}} + q\beta_{\text{intrinsic}}} \\
&= (-q\beta_{\text{intrinsic}}) \times \frac{2.303kT}{2.303kTC_{\text{differential}} + q\beta_{\text{intrinsic}}} \\
&= (-q\beta_{\text{intrinsic}}) \times \frac{2.303kT}{q\beta_{\text{intrinsic}}\left[(2.303kTC_{\text{differential}}/q\beta_{\text{intrinsic}}) + 1\right]} \\
&= -\frac{2.303kT}{\left[(2.303kTC_{\text{differential}}/q\beta_{\text{intrinsic}})\right] + 1} \\
&= -\left(\frac{2.303kT}{q}\right) \frac{1}{\left[(2.303kTC_{\text{differential}}/q^2\beta_{\text{intrinsic}})\right] + 1}
\end{aligned} \tag{8.49}$$

The limiting value of the oxide surface sensitivity at $T = 298$ K is

$$\frac{\partial \psi_0}{\partial pH_B} = -59.2 \text{ mV/pH} \tag{8.50}$$

The change of ψ_0 with pH_B is translated into a change in conductance ΔG in the nano-ISFET produced by an electric field across the oxide layer, enabling the experimental determination of the variation of ΔpH_B with $\Delta\psi_0$.

As the changes in pH of the solution affect the electrostatic potential of the surface, and in turn electrostatic potential influences the carrier concentration in the nanowire, it follows that the sensitivity of the nanowire to pH changes should scale with inverse of the nanowire diameter.

6. SUBTHRESHOLD REGIME AS THE OPTIMAL SENSITIVITY REGIME OF NANOWIRE BIOSENSORS

In the equation for sensitivity, ΔG represents the portion of the NW volume gated by surface charges, while G signifies the whole volume of the nanowire. Thus sensitivity is essentially a fraction equal to the ratio of two NW volumes. The maximum value of sensitivity is reached when this ratio has the highest possible value of unity. This highest value corresponds to the situation when the whole volume of the nanowire is gated. Such gating is not realized at typical nanowire doping densities because the Debye screening length in silicon $(\lambda_D)_{Si}$ is ~1–2 nm at these carrier concentrations, while the nanowires used have radii ~10 nm. Complete gating of the nanowire is possible at much lower carrier concentrations because at these concentrations, $(\lambda_D)_{Si} >> R$. Hence, sensitivity is expected to increase at lower carrier concentrations, which are achieved when the nanowire is operated in depletion mode (Figure 8.10).

Gao et al. (2010) carried out extensive pH sensing experiments in the linear regime, near threshold voltage and in the subthreshold regime. They found that in the subthreshold regime, the device showed a much larger percentage change in conductance with pH variation than in other regimes of operation. Table 8.4 presents a comparative study of the linear and subthreshold regimes. Gao et al. (2010) argued convincingly to prove that the lowest detection limit of the NW sensor is achieved in the subthreshold regime. Following their approach, for a surface potential $(\Delta\phi)_{SiO_2}$ at the SiO_2/electrolyte boundary, the charge detected is written as

$$\Delta\theta = C_{total}(\Delta\phi)_{SiO_2} \tag{8.51}$$

where C_{total} is the total capacitance constituted by the surface charge and the nanowire/oxide/electrolyte layers. The capacitance C_{total} consists of three components: (1) nanowire capacitance C_{NW}, (2) silicon dioxide capacitance C_{SiO_2}, and (3) capacitance of the electrical double layer in the electrolyte (C_{DL}). Among these

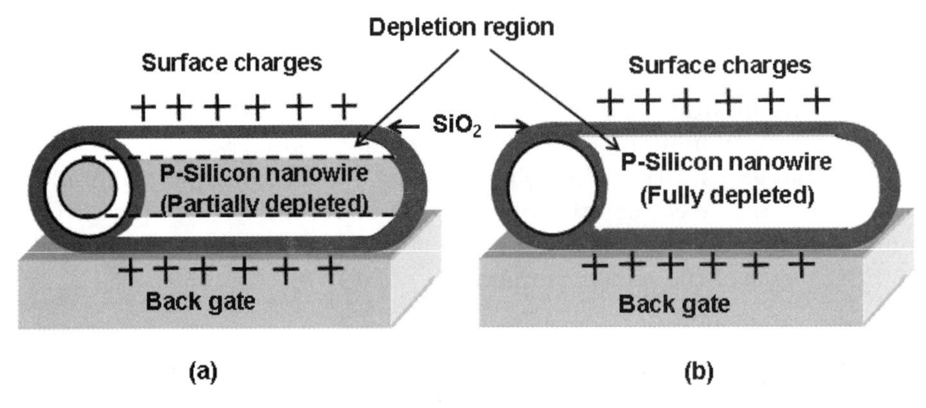

Figure 8.10. Nanowire operational domain: (a) linear domain in which the depletion region extends to a small depth below the surface; (b) subthreshold domain in which the depletion region spreads to the whole diameter of the nanowire. .

Table 8.4. Linear versus subthreshold-regime operation of nanowire

Sl. No.	Linear-regime operation	Subthreshold-regime operation
1.	High carrier-concentration regime.	Low carrier-concentration regime.
2.	Conductance G is directly proportional to gate voltage V_G.	Conductance G varies exponentially with gate voltage V_G.
3.	Debye screening length $(\lambda_D)_{Si}$ is << radius R of the nanowire.	Debye screening length $(\lambda_D)_{Si}$ is >> radius R of the nanowire.
4.	By field effect, the surface charges induce band bending within a region of depth ~ screening length $(\lambda_D)_{Si}$.	The whole nanowire is depleted of free charge carriers; hence it is fully utilized for sensing.
5.	Surface charges induce carrier depletion/ enhancement up to a depth ~ screening length.	Surface charges gate the whole nanowire, fully utilizing the nanowire surface for gating and volume for carrier depletion.
6.	Sensitivity is lower than in subthreshold region.	Maximum-sensitivity regime of operation.

components, the two capacitances C_{NW} and C_{SiO_2} are connected in series, giving a capacitance

$$C_{Series} = \left(\frac{1}{C_{NW}} + \frac{1}{C_{SiO_2}}\right)^{-1} \tag{8.52}$$

The capacitance C_{Series} is connected in parallel with C_{DL}, yielding a resulting capacitance

$$C_{total} = \left(\frac{1}{C_{NW}} + \frac{1}{C_{SiO_2}}\right)^{-1} + C_{DL} \tag{8.53}$$

From Eq. (8.51), the minimum detectable charge is

$$(\Delta\theta)_{minimum} = (C_{total})_{minimum} \times \left\{(\Delta\phi)_{SiO_2}\right\}_{minimum} \tag{8.54}$$

To detect $(\Delta\theta)_{minimum}$, C_{total} is reduced to $(C_{total})_{minimum}$, and for this purpose, C_{NW} and C_{DL} are decreased. In the subthreshold region, $C_{NW} \rightarrow 0$; hence, the smallest charge detection will be possible in the subthreshold regime. $\left\{(\Delta\phi)_{SiO_2}\right\}_{minimum}$ is determined by the electrochemistry at the oxide/electrolyte interface and the noise characteristics of the nanowire. The Debye screening length in the electrolyte is

$$\lambda_{\text{Electrolyte}} \propto \frac{1}{\sqrt{\text{ionic strength of solution}}} \tag{8.55}$$

As the ionic strength decreases, the screening length will increase. On the whole, device operation in the subthreshold region and in a low-ionic-strength electrolyte is prescribed as the necessary condition for reaching the limiting value in which very few charges are detectable.

7. EFFECTIVE CAPACITANCE MODEL FOR APPARENT SURPASSING OF THE NERNST LIMIT BY SENSITIVITY OF THE DUAL-GATE NW SENSOR

In a dual-gate NW FET sensor, the carrier density is modulated by applying electrostatic potentials to the back gate (V_{BG}) underneath the nanowire, and to the front or liquid gate (V_{LG}) above the nanowire. The latter potential is applied through the electrolyte by connecting a platinum wire immersed in the same and connected to the power source. Such simultaneous front- and back-side gating helps in configuring the device to different operating points, enabling some control over sensitivity.

Knopfmacher and co-workers (2010, 2011) found that the apparent sensitivity of a dual-gate NW sensor to pH can be enormously high, ~220 mV/pH, which is well above the ideal Nernst limit. For interpretation of these anomalous results, they proposed an effective capacitance network model including both gates. This network is shown in Figure 8.11. In this circuit, the back-gate capacitance is composed of a series combination of electrical double-layer capacitance C_{DL} in the electrolyte and the silicon dioxide capacitance C_{SiO_2}. In addition, the circuit includes the nanowire capacitance C_{NW}. The capacitor C_{Contact} is included to represent the contact capacitance of the nanowire source and drain contacts.

When the change in conductance is measured with respect to the liquid gate, sensitivity values below the Nernst limit are obtained, as expected. But when the pH-induced shifts are measured relative to the back gate, sensitivity limits above the Nernst limit are noticed. For calculating the shifts δV_{BG} per pH change, they found the responses $\delta Q/\delta V_{\text{BG}}$ and $\delta Q/\delta V_{\text{LG}}$ (where Q denotes the screening charge of the nanowire), and related them to get

$$\frac{\delta V_{\text{LG}}}{\delta V_{\text{BG}}} = \frac{C_{\text{BG}}}{C_{\text{LG}}} \tag{8.56}$$

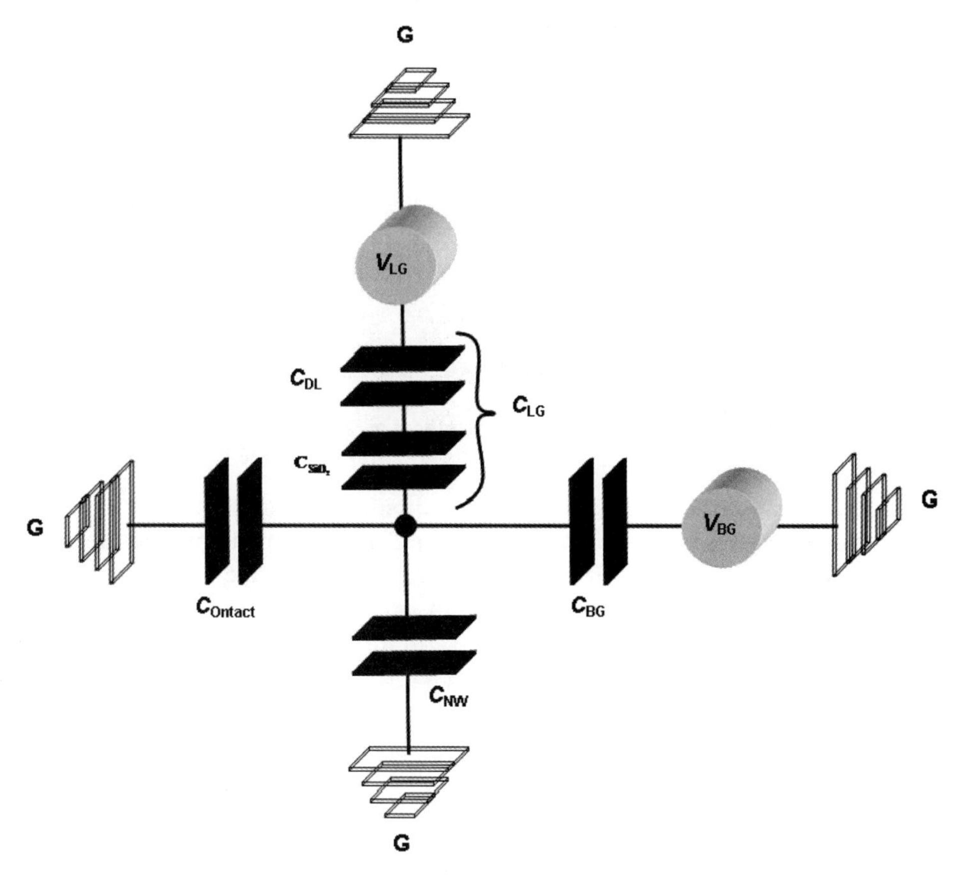

Figure 8.11. Effective capacitance network for interpreting the surprisingly high sensitivity of silicon nanowire with respect to the back gate, beyond the theoretical Nernst limit. The model shows the back-gate capacitor, the liquid-gate capacitance components (double-layer and nanowire capacitors), and the contact capacitor. (Idea from Knopfmacher et al. 2010.)

Remembering that the liquid-gate capacitance has the series-connected components C_{DL} and C_{SiO_2}, and at the high ion concentrations used in buffer solutions, $C_{DL} >> C_{SiO_2}$, the capacitance C_{SiO_2} will have the predominant effect, so that, in practice,

$$\frac{\delta V_{LG}}{\delta V_{BG}} = \frac{C_{BG}}{C_{DL,\,SiO_2}} \approx \frac{C_{BG}}{C_{SiO_2}} \tag{8.57}$$

and voltage division takes place in the ratio C_{SiO_2}/C_{BG}. If C_{BG} is $< C_{SiO_2}$, the pH-induced back-gate voltage shift is enhanced. Under these circumstances, an apparent amplifying action is produced, and the pH sensitivity overshoots the Nernst limit. However, if C_{BG} is $> C_{SiO_2}$, the pH-induced shift is diminished.

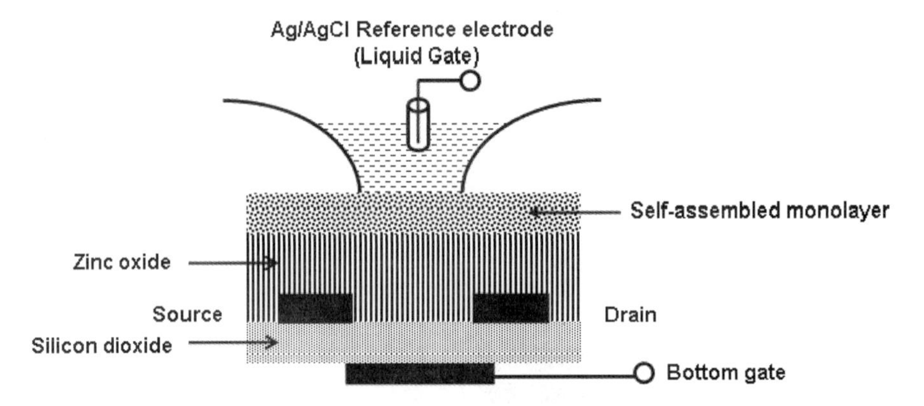

Figure 8.12. Dual-gate ZnO ISFET. (Idea from Spijkmanet al. 2011.)

Spijkman et al. (2011) demonstrated a transducer based on a ZnO dual-gate FET that breached the Nernst boundary of 59.2 mV/pH by a spectacular margin (Figure 8.12). A high top capacitance is necessary to increase the capacitive coupling C_{top} / C_{bottom} between the top and bottom gates. A large top-gate capacitance was obtained by applying a self-assembled monolayer (SAM) of octadecylphosphonic acid on top of the semiconducting ZnO, thereby passivating this film. The resulting sensitivity of 2250 mV/pH was orders of magnitude larger than the Nernstian response of 59.2 mV/pH.

8. TUNNEL FIELD-EFFECT TRANSISTOR CONCEPT

Sarkar and Banerjee (2012) proposed the concept of a hitherto-unrealized, tunnel field-effect transistor (TFET) as an ultrasensitive device (Figure 8.13a). This device is based on a fundamentally different current injection mechanism than the previous notion of FET sensors. In this device, the current injection takes place under the influence of biomolecular charges attracted toward the gate, but unlike the regular FETs, gating action modulates the band-to-band tunneling barrier, enabling the flow of current from the source to the channel. Initially, when no biomolecules are conjugated with the receptors on the gate, the tunneling barrier is high, stopping the current flow from the source to the channel. However, upon capturing biomolecules by the receptors, the bands in the channel bend downwards. Consequently, the tunneling barrier is lowered and the tunneling current is increased. The energy-band diagrams for the tunnel FET in both the conditions, i.e., before and after biomolecular capture, are depicted in Figures 8.13b and 8.13c.

From analytical considerations, a formula was derived for sensitivity S_n defined as

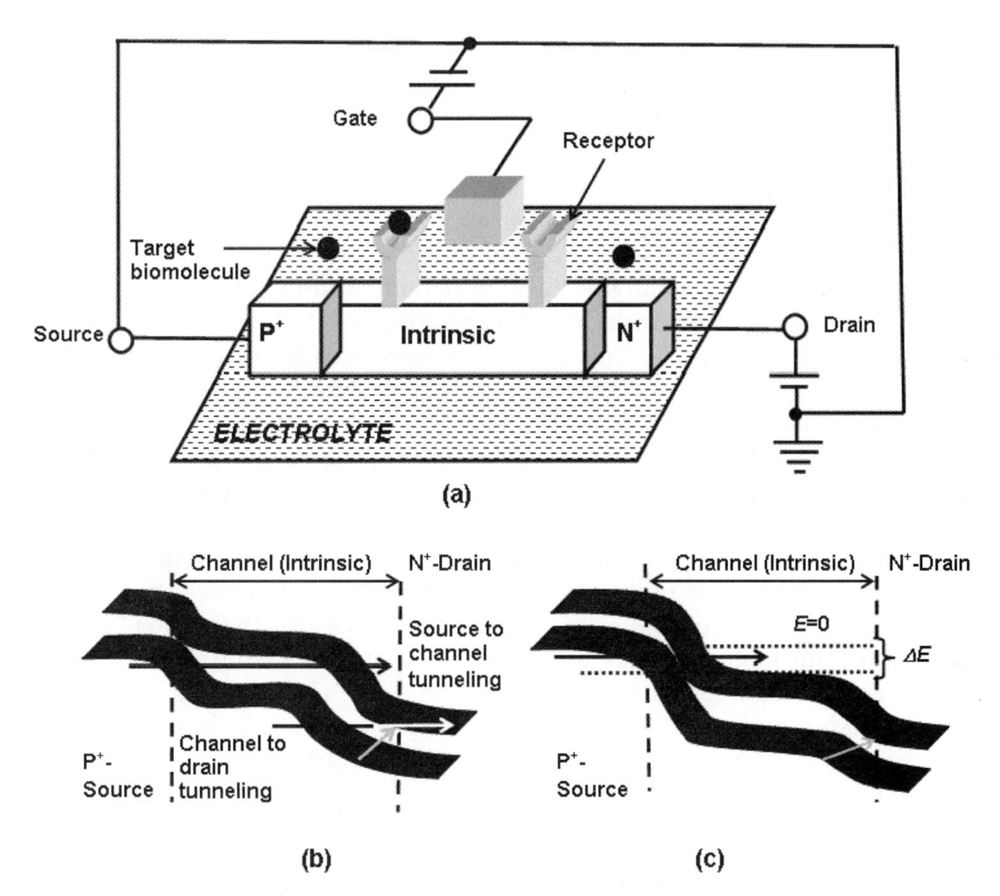

Figure 8.13. Structure of tunnel FET biosensor, and its energy-band diagram representation. (a) Reverse-biased P^+–intrinsic layer–N^+ structure of the sensor; the liquid gate immersed in the solution controls the operational domain of the device. (b) Band diagram during OFF state, with the large barrier between the source and channel preventing tunneling current flow. (c) Band diagram in ON state, marked by downward bending of the bands, barrier reduction, and increased current flow by tunneling from source to channel. (Idea from Deblina and Sarkar 2012.)

$$S_n = \frac{I_{\text{BTBT}}\left(\phi_0 + \phi_{\text{bio}}\right) - I_{\text{BTBT}}\left(\phi_0\right)}{I_{\text{BTBT}}\left(\phi_0\right)} \tag{8.58}$$

where I_{BTBT} $(\phi_0+\phi_{\text{bio}})$ is the barrier-to-barrier tunneling current after biomolecule capture and I_{BTBT} (ϕ_0) is the same current before capture.

The sensitivity formula is

$$S_n = \exp\left[\frac{\pi\sqrt{2m^*}qE_G^{\,1.5}\lambda\phi_{\text{bio}}}{\hbar\left(2\phi_0 - E_G\right)\left(2\phi_0 + 2\phi_{\text{bio}} - E_G\right)}\right] \times \left(1 + \frac{\phi_{\text{bio}}}{\phi_{\text{bio}} - E_G}\right) - 1 \tag{8.59}$$

where the symbols have the following meanings: ϕ_0 = potential on the SiO_2 surface at the beginning of the experiment, prior to biomolecular attachment; ϕ_{bio} = potential on the oxide surface after the attachment of biomolecules; λ = natural length scale; E_G = energy band gap of the semiconductor material in eV; m^* = effective mass of the carriers; and $\hbar = h/2\pi$, where h is Planck's constant. The analytical equation provides physical insights governing the sensor operation. If ΔE denotes the energy difference in the band diagram from the valence band of the source to the point of complete flattening of the conduction band in the channel, for $\Delta E \geq 0$, the sensitivity increases with decreasing ϕ_0 because the rate of rise of current with gate voltage is larger for smaller ΔE and therefore lower ϕ_0 values. Further, the low ΔE value suggests subthreshold mode of operation of the device, which means that higher sensitivity is a characteristic feature of this domain. Another vital inference is that the sensitivity will be more for larger band gap of the semiconducting material used in TFET fabrication. This happens because the initial current of the device before capturing biomolecules decreases. The researchers pointed out that the subthreshold regime is also the optimal sensing regime for planar micro-ISFETs. The planar ISFETs suffer from the drawback that the minimum achievable subthreshold swing is restricted. However, the tunnel FET is not afflicted with this limitation, due to the Fermi tail cutting by the E_G. For the same reason, these researchers indicated that tunnel FETs will provide much faster response than planar ISFETs. Over four orders of magnitude increase in sensitivity together with one order of magnitude acceleration of response behavior is theoretically predicted.

9. ROLE OF NANOPARTICLES IN ISFET GATE FUNCTIONALIZATION

9.1. SUPPORTIVE ROLE OF NANOPARTICLES

In the preceding sections, the various physicochemical phenomena explaining the operation of nanostructured channel ISFETs were discussed. ISFET-based biosensors have long been realized by immobilizing various enzymatic membranes on the gate dielectric of a planar ISFET device (Shipway and Milner 2001; Xu et al. 2005; Gun et al. 2008; Vijayalakshmi et al. 2008). These enzymes act as biocatalysts to speed up a reaction which releases protons and thus the biosensor becomes specific to the particular biomolecules whose reaction is promoted by the enzyme concerned. The enzymes are highly selective in action, imparting selectivity to the biosensor. Several instances can be cited in recent literature in which nanoparticles such as gold, platinum, manganese dioxide, silica, etc., have been added to the enzymes.

The obvious advantage of nanoparticle-based enzyme preparations originates from the large surface area-to-volume ratio of the added nanoparticles. The large specific surface area and high surface energy of nanoparticles enables them to adsorb the biomolecules strongly. Carrying of charges by most of the nanoparticles helps them to electrostatically adsorb biomolecules with different charges.

Luo et al. (2004) fabricated a glucose-sensitive ENFET by modification of the gate surface of an ISFET with SiO_2 nanoparticles and glucose oxidase (GOD). In contrast to the sensor without SiO_2 nanoparticles, the nanoparticle-based sensor showed enhanced sensitivity and extended lifetime. This implies that SiO_2 nanoparticles provided a biocompatible environment for the enzyme and improved its activity. Additionally, they prevented the immobilized enzyme from leakage.

The use of SiO_2 nanoparticles as an immobilization matrix of GOD to the gate surface of the ENFET greatly ameliorated the response of the biosensor and ensured its longevity. Firsty, SiO_2 nanoparticles have a large surface area, due to which they adsorb the GOD intensively and prevent the enzyme from leakage. Second, owing to the biocompatibility of SiO_2 nanoparticles, they serve as a favorable environment for GOD to maintain its bioactivity. The linear ranges of the ISFET without nanoparticles and the ISFET with nanoparticles were 0.10–1.70 and 0.05–1.80 mM, respectively, and the corresponding detection limits were 0.05 and 0.025 mM, respectively. The nanoparticle-based biosensor also exhibited acceptable long-term stability. When stored in dry state at 4°C, the response of a newly prepared nanoparticle ISFET showed a drop to ~70% of its original value during the first 9 days. This may be attributed to the partial leakage of the GOD that adsorbed on the fringe of the sensitive area. Thereafter, the sensor remained stable for more than 5 weeks. A similar phenomenon was also observed for the ISFET without nanoparticles, but it showed inferior stability (~3 weeks) compared with the ISFET with nanoparticles. Since the nanoparticle glucose biosensor furnished appreciably larger response, it was usable for glucose detection, by recalibration at intervals, for at least a month.

Yao et al. (2008) fabricated an ISFET glucose biosensor by immobilizing a nanocomposite film containing dendrimer-encapsulated Pt nanoparticles and glucose oxidase (GOx) via a layer-by-layer self-assembly method. The glucose-sensitive ISFET developed showed higher sensitivity and extended lifetime compared with conventional ones. The fabricated sensor had a linear range of 0.25–2.0 mM and a detection limit of ~0.15 mM. Yao et al. (2008) found that the response of the sensor in which the Pt-DENs/GOx nanobiocomposite film was employed was superior to that of the sensor which did not contain the nanocomposite. This showed that the nanobiocomposite film helped in improving the sensor capability. They classified the nanobiocomposite film as a conductive material, a supporting platform, and a biomolecular material in accordance with the nature of Pt dendrimers and GOx. While the glucose oxidase enzyme was responsible for the selectivity of the device to a given analyte, the Pt nanoparticles acted as the conductive materials encapsulated in the dendrimers' internal porosity. Metal–glucose oxidase interactions were avoided, and replaced by biocompatible polymer–glucose oxidase interactions. This afforded more flexibility in the deposition and dispersion of metal nanoparticles in the film to provide special nano effects along with the possibility of reducing enzyme denaturing.

It is observed that the adsorption of biomolecules directly onto naked surfaces of bulk materials frequently leads to their denaturation and loss of bioactivity. But the adsorption of such biomolecules on the surfaces of nanoparticles aids in retention of their bioactivity because of the biocompatibility of nanoparticles (Luo et al. 2006). Therefore, the immobilization of biomolecules with nanoparticles effectively promotes the stability and maintains the activity of biomolecules.

9.2. DIRECT REACTANT ROLE OF NANOPARTICLES

Sometimes nanoparticles exhibit remarkably different behavior from the bulk particles, leading to unexpected results. A conspicuous example is the use of MnO_2 nanoparticles in glucose ENFETs, in which the nano-MnO_2 ENFET was found to show opposite behavior to an ENFET fabricated without MnO_2 nanoparticles (Luo et al. 2004). In the nano-MnO_2 ENFET, the local pH on the enzyme layer registered an increase in pH with glucose concentration, i.e., shifting toward the alkaline side instead of the acidic side, as is usually observed in glucose ENFETs as the glucose concentration rises. In a regular glucose ENFET which does not have such nanoparticles, glucose forms glunolactone in the presence of water in the solution and oxygen in the atmosphere; this gluconolactone subsequently dissociates to liberate protons that are detected by the ISFET. The higher the glucose concentration, the larger is the number of protons liberated from gluconolactone, and therefore the less is the local pH of the solution. This is the sensing principle of a glucose-ENFET without the addition of MnO_2 nanoparticles to the glucose oxidase enzyme (Luo et al. 2004):

$$\beta\text{-D-Glucose} + O_2 \rightarrow \text{D-glucono-}\delta\text{-lactone} + H_2O_2 \quad \text{(in the presence of glucose oxidase)} \qquad (8.60)$$

$$\text{D-Glucono-}\delta\text{-lactone} + H_2O \rightarrow \text{D-gluconate} + H^+ \qquad (8.61)$$

To explain the behavior reversal noticed in the nano-MnO_2 ISFET, the difference in chemical reactivity between bulk MnO_2 particles and MnO_2 nanoparticles was investigated. Although bulk MnO_2 is known to act as a catalytic agent for the decomposition reaction of H_2O_2 into water and oxygen, it was found that nano-MnO_2 particles could react directly with H_2O_2. This participation of nano-MnO_2 in the reaction became evident when H_2O_2 was added to colloidal MnO_2 solution. This resulted in the formation of oxygen bubbles together with the sedimentation of a brown precipitate. A simultaneous increase in pH by as much as 3.5 pH units was recorded with 15 mM H_2O_2 solution. The sedimentation was caused by aggregation or clustering of MnO_2 nanoparticles. Oxidation of H_2O_2 to produce oxygen and Mn^{2+} was confirmed by suitable chemical reactions and supplemented by absorption spectrum studies.

The above observations were analyzed, and the sensing mechanism of the nano-MnO_2 ISFET was proposed in terms of the reactions taking place between MnO_2 and H_2O_2, yielding MnOOH and oxygen. Essentially, MnO_2 nanoparticles serve as an oxidizing agent, playing the role of reactant in the reaction and *not* that of a catalyst. The revised glucose detection reactions are (Luo et al. 2004)

$$\beta\text{-D-glucose} + O_2 + H_2O \rightarrow \text{D-gluconate} + H_2O_2 + H^+$$
(in the presence of glucose oxidase)

$$2MnO_2 + H_2O_2 \rightarrow 2MnOOH + O_2 \tag{8.62}$$

$$2MnOOH + 4H^+ + H_2O_2 \rightarrow 2Mn^{2+} + 4H_2O + O_2 \tag{8.63}$$

$$2MnOOH + 2H^+ \rightarrow MnO_2 + Mn^{2+} + 2H_2O \tag{8.64}$$

The net reaction is

$$\beta\text{-D-glucose} + MnO_2 + H^+ \rightarrow Mn^{2+} + \text{D-gluconate} + H_2O \tag{8.65}$$

This reaction seizes a proton. The consumption of a proton through this reaction decreases the local hydrogen ion concentration and thereby increases the local pH of the solution. This increase in pH, in contrast to the decrease in pH taking place in an ISFET without MnO_2 nanoparticles, explains the reversal of response of the nano-MnO_2 ISFET.

The oxygen released during the decomposition of H_2O_2 also contributes to improving the performance of this ISFET. It is recycled and reused in glucose conversion to gluconolactone. This re-availability of oxygen extends the upper dynamic range of the nano-MnO_2 sensor. Consequently, the nano-MnO_2 ISFET has a higher upper limit of dynamic range (3.5 mM) than the usual glucose ENFET (2.0 mM).

The revelations of higher chemical reactivity of nanoparticles are not astonishing when one recalls that nanoparticles are more active than bulk particles by virtue of their higher surface energy. A similar biosensor employing MnO_2 nanoparticles with lactate oxidase enzyme was reported to give 50 times superior response than one without MnO_2 nanoparticles (Xu et al. 2005). Another sensor, for ascorbic acid detection, utilized this nanoparticle effect to enhance the responsivity (Luo et al. 2004). It was therefore conjectured that the unique properties of nanoparticles such as PbO_2, CeO_2, etc., could be beneficially applied to tailor the sensitivity and detection range of electrochemical biosensors, opening a promising research field.

10. NEURON-CNT (CARBON NANOTUBE) ISFET JUNCTION MODELING

Nanomaterials in the form of carbon nanotubes are anchored to the gates of planar ISFETs to serve as interfaces with neurons. By this mechanism, the efficacy of signal transmission from neurons to the ISFET increases appreciably, enabling efficient extracellular measurement of the electrophysiological activity of neurons. For a clear understanding and more accurate interpretation of signals, models and simulation tools are necessary.

Mossobrio et al. (2008) modeled the response of a system comprising neurons interacting with carbon nanotubes arranged vertically over the gate insulator of an ISFET. Mossobrio et al. (2011) elaborated the development of models of the neuron, ISFET, CNT, and neuroelectronic junction, the modification of these models, and their implementation into the circuit simulation program HSPICE.

In the *neuron model,* the Na^+ channels are modeled by four MOSFETs and two capacitors; K^+ channels are modeled by two MOSFETs and one capacitor; and the neuron membrane by one capacitor. Control over the activation/inactivation time constants of the Na^+ and K^+ channels is provided by acting on the gate bias voltages of the MOSFETs.

In the *SWCNT model,* each SWCNT comprises lumped resistors representing the intrinsic ballistic resistance of the SWCNT and contact resistance between the SWCNT and on-chip metal. The ohmic resistance of the SWCNT is considered in a distributed resistance, while the resistance between neighboring SWCNTs is accounted for by a coupling resistance. The three capacitances, viz., the quantum capacitance of the SWCNT, the electrostatic capacitance to ground, and the capacitance between SWCNTs, are represented by three separate capacitors (per unit length). Owing to the high intrinsic ballistic resistance of an isolated CNT, a bundle of parallel-connected CNTs is used to supply a high drive current.

In the *ISFET model,* there are two fully uncoupled stages: an *electrochemical stage* for the electrolyte-insulator interface, based on the electrical double-layer and site-binding theories; and an *electronic stage* working according to the MOSFET theory.

In the electrical equivalent circuits for neuron–ISFET and neuron–CNT–ISFET junctions, Figure 8.14, R_{sealing} is the seal resistance between the cell and the recording sensitive area of the ISFET device to which the neuron cell is affixed. The spreading resistance $R_{\text{spreading}}$ accounts for the signal loss occurring during transmission over the distance between the neuron and the ISFET gate insulator surface and/or the top surface of the SWCNT bundle. C_{hd} (neuron membrane-to-electrolyte capacitance) is the series combination of the Helmholtz layer capacitance and the Gouy-Chapman or diffuse layer capacitance. Z_{SWCNT} is the impedance of the isolated SWCNT, and Z_{Coupling} is the coupling impedance between two SWCNTs.

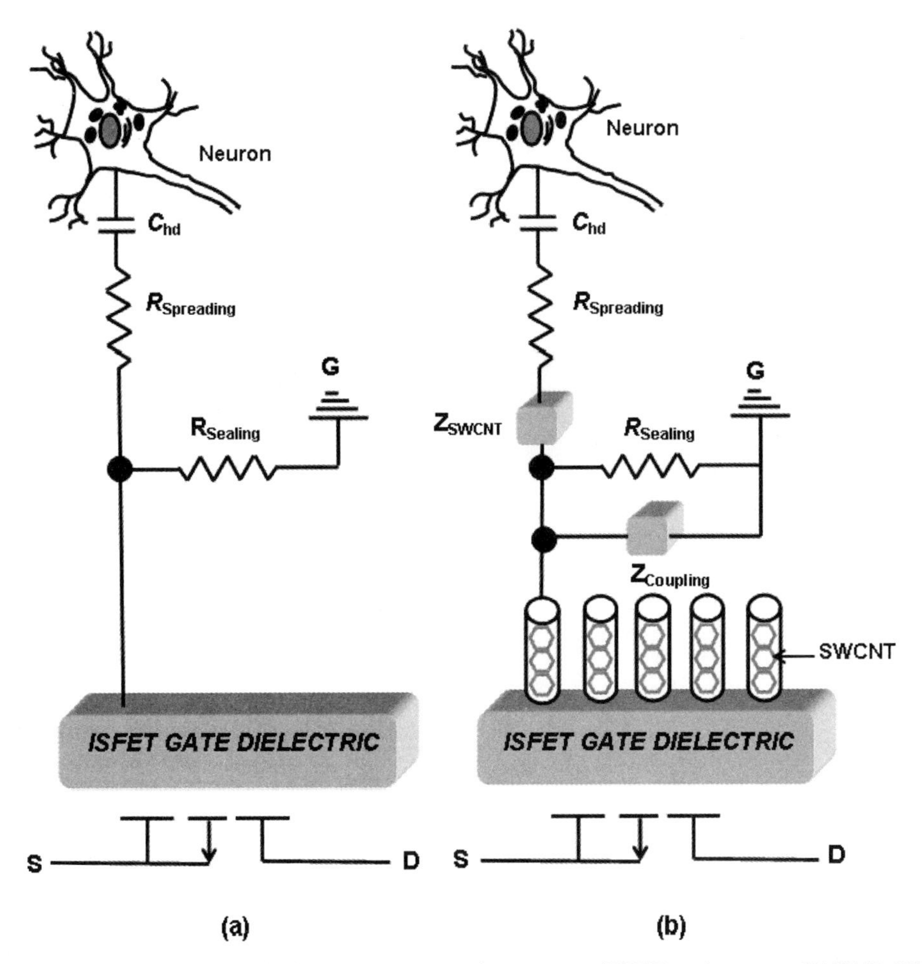

Figure 8.14. Equivalent circuit diagrams representing neuron–ISFET and neuron–SWCNT–ISFET junctions. (Idea from Massobrio et al. 2008, 20011.)

Simulations of the behavior of the above two circuit configurations, one without SWCNTs and the other with SWCNTS, were performed using HSPICE. The neuronal electrical activity was simulated by varying the neuro-electronic junction parameters such as the seal resistance, double-layer capacitance, and adhesion conditions. Simulation studies were also conducted with respect to SWCNT properties, e.g., diameter, length, number in the bundle, bundle geometry, distance from neurons, etc. The simulations provided useful information about the shape of the recorded neuronal signals under limiting conditions of feeble and sturdy coupling. The simulation results strengthened the belief that SWCNTs are useful devices for improving neural signal transference processes. The SWCNTs were found to affect both the amplitude and the shape of the recorded signals.

11. CONCLUSIONS AND PERSPECTIVES

Incorporation of nanoscale effects in ISFET devices by two approaches has been reviewed (Figure 8.15). The first approach is the downscaling of the channel to nano dimensions. The second approach is the embedding of nanoparticles in the

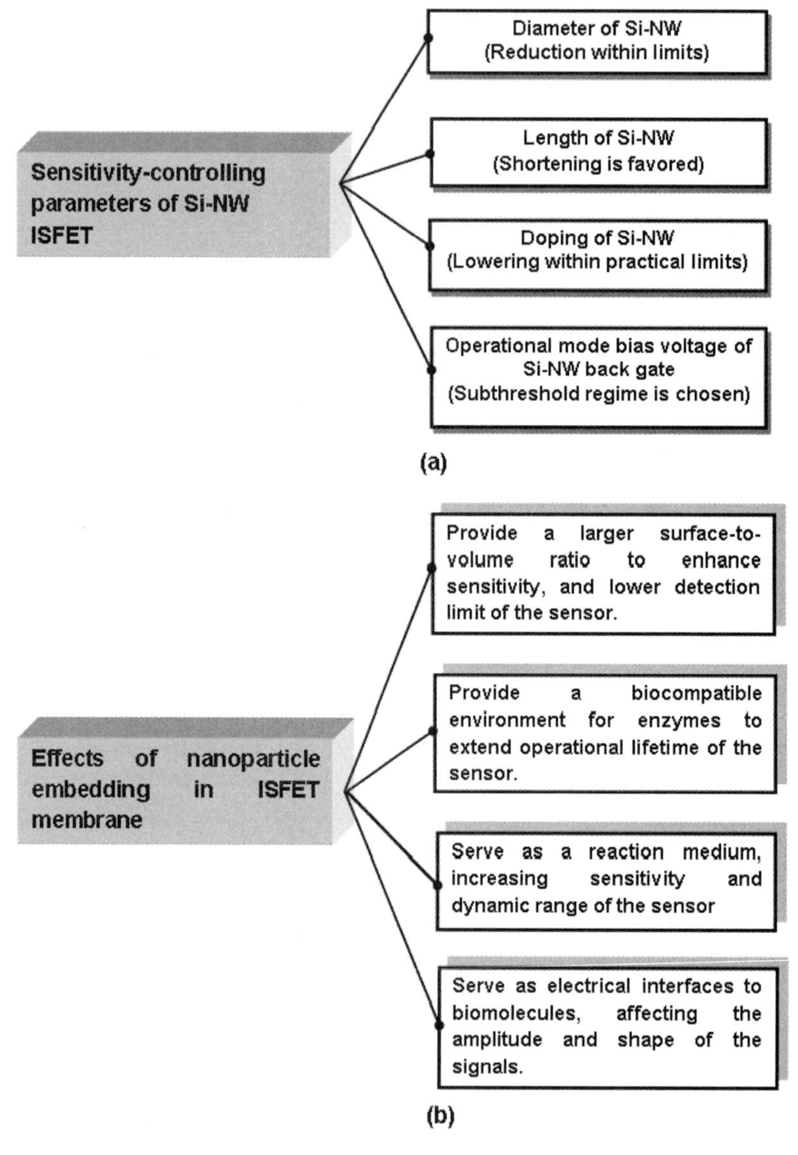

Figure 8.15. (a) Main design considerations of nanoISFETs, and (b) effects of nanoparticles in ISFET membrane.

membranes on the gate of micro ISFET devices. Various models describing the behavior of these new devices have been surveyed. These models help not only in understanding the transduction mechanisms of ISFET-based nanosensors for chemical and biological applications, they also serve as a pathway guiding the fabrication of novel devices. At the bottom of all these efforts lies the unending quest for methods of improving ISFET performance characteristics. These exploratory endeavors continue, giving birth to newer ideas and concepts.

DEDICATION

To my late father, Shri Amarnath Khanna, for nurturing my educational career.

ACKNOWLEDGMENTS

Director, CEERI, Pilani, for encouragement, motivation and guidance.

REFERENCES

Attard P. (1996) Electrolytes and the electric double layer. *Adv. Chem. Phys.* **92**, 1–159.

Bergveld P. (2003) ISFET, Theory and practice. In: *Proceedings of the IEEE Sensor Conference*, Toronto, Ontario, Oct., 1–26, http://ewh.ieee.org/tc/sensors/Tutorials/ISFET-Bergveld.pdf.

Carlen E.T. and van den Berg A. (2007) Nanowire electrochemical sensors: Can we live without labels. *Lab Chip* **7**, 19–23.

Chen S., Bomer J.G., Carlen E.T., and van den Berg A. (2011) Al_2O_3/silicon nano ISFET with near ideal Nernstian response. *Nano Lett.* **11**, 2334–2341. DOI: 10.1021/nl200623n

Chiesa M., Cardenas P.P., Oton F., Martinez J., Mas-Torrent M., Garcia F., Alonso J.C., Rovira C., and Garcia R. (2012) Detection of the early stage of recombinational DNA repair by silicon nanowire transistors. *Nano Lett.* **12**, 1275-1281. DOI: 10.1021/nl2037547

Dong Z., Wejinyaa U.C., and Chalamalasetty S.N.S. (2012) Development of CNT-ISFET based pH sensing system using atomic force microscopy. *Sens. Actuators A* **173**, 293–301. DOI: 10.1016/j.sna.2011.10.029

Dong Z., Wejinya U.C., Yu H., and Elhajj I.H. (2009) Design, fabrication and testing of CNT-based ISFET for nano pH sensor application: A preliminary study. In: *Proceedings of 2009 IEEE/ASME International Conference on Advanced Intelligent Mechatronics*, Singapore, July 14–17, 1556–1561.

Fan Z. and Lu J.G. (2006) Chemical sensing with ZnO nanowire field-effect transistor. *IEEE Trans. Nanotechnol.* **5**(4), 393–396. DOI: 10.1109/TNANO.2006.877428

Fan Z. and Lu J.G. (2005) Gate-refreshable nanowire chemical sensors. *Appl. Phys. Lett.* **86**, 123510. DOI: 10.1063/1.1883715

Gao X.P.A., Zheng G., and Lieber C.M. (2010) Subthreshold regime has the optimal sensitivity for nanowire FET biosensors. *Nano Lett.* **10**, 547–552. DOI: 10.1021/nl9034219

Gun J., Schöning M.J., Abouzar M.H., Poghossian A., and Katz E. (2008) Field-effect nanoparticle-based glucose sensor on a chip: Amplification effect of coimmobilized redox species. *Electroanalysis* **20**(16), 1748–1753. DOI: 10.1002/elan.200804255

Heitzinger C. and Klimeck G. (2007) Computational aspects of the three-dimensional feature-scale simulation of silicon nanowire field-effect sensors for DNA detection. Birck and NCN Publications, Paper 440, http://docs.lib.purdue.edu/nanopub/440.

Heitzinger C., Kennell R., Klimeck G., Mauser N., McLennan M., and Ringhofer C. (2008) Modeling and simulation of field-effect biosensors (BioFETs) and their deployment on the nanoHUB. *J. Phys.: Conf. Ser.* **107**, 012004. DOI: 10.1088/1742-6596/107/1/012004

Katz E., Willner I., and Wang J. (2004) Electroanalytical and bioelectroanalytical systems based on metal and semiconductor nanoparticles. *Electroanalysis* **16**(1–2), 19–44. DOI: 10.1002/elan.200302930

Kim C.-H., Jung C., Lee K.-B., Park H.G., and Choi Y.-K. (2011) Label-free DNA detection with a nanogap embedded complementary metal oxide semiconductor. *Nanotechnology* **22**, 135502. DOI: 10.1088/0957-4484/22/13/135502

Kim S., Rim T., Kim K., Lee U., Baek E., Lee H., Baek C.-K., Meyyappan M., Deen M.J., and Lee J.-S. (2011) Silicon nanowire ion-sensitive field effect transistor with integrated Ag/AgCl electrode: pH sensing and noise characteristics. *Analyst* **136**, 5012–5016. DOI: 10.1039/C1AN15568G

Knopfmacher O.S. (2011) *Sensing with Silicon Nanowire Field-Effect Transistors.* Inaugural dissertation, Uni Basel, Basel.

Knopfmacher O.S., Tarasov A., Fu W., Wipf M., Niesen B., Calame M., and Schönenberger C. (2010) Nernst limit in dual-gated Si-nanowire FET sensors. *Nano Lett.* **10**, 2268–2274. DOI: 10.1021/nl100892y|

Lee C.-S., Kim S.K., and Kim M. (2009) Ion-sensitive field-effect transistor for biological sensing. *Sensors* **9**, 7111–7131. DOI: 10.3390/s90907111

Lee D. and Cui T. (2010) Low-cost, transparent, and flexible single-walled carbon nanotube nanocomposite based ion-sensitive field-effect transistors for pH/glucose sensing. *Biosens. Bioelectron.* **25**, 2259–2264. DOI: 10.1016/j.bios.2010.03.003

Lee J. and Shin M. (2009) Effects of pH and ion concentration in a phosphate buffer solution on the sensitivity of silicon nanowire BioFETs. *J. Korean Phys. Soc.* **55**(4), 1621–1625.

Lin Y.-T., Yu Y.-H., Chen Y., Zhang G.-J., Zhu S.-Y., Yang C.-M., Lu K.-Y., and Lai C.-S. (2011) Vertical silicon nanowires with atomic layer deposition with HfO_2 membrane for pH sensing application. *J. Mech. Med. Biol.* **11**(5), 959–966. DOI: 10.1142/S0219519411004897

Liu Y., Guo Q., Wang S., and Hu W. (2012) Electrokinetic effects on detection time of nanowire biosensor. *Appl. Phys. Lett.* **100**, 153502. DOI: 10.1063/1.3701721

Liu Y., Lilja K., Heitzinger C., and Dutton R.W. (2008) Overcoming the screening-induced performance limits of nanowire biosensors: A simulation study on the effect of electro-diffusion flow. In: *Proceedings of IEDM 2008: IEEE International Electron Devices Meeting,* 15–17 Dec. 2008, San Francisco, CA, 491–494.

Luo X.-L., Xu J.-J., Zhao W., and Chen H.-Y. (2004) A novel glucose ENFET based on the special reactivity of MnO_2 nanoparticles. *Biosens. Bioelectron.* **19**, 1295–1300. DOI: 10.1016/j.bios.2003.11.019

Luo X.-L., Xu J.-J., Zhao W., and Chen H.-Y. (2004) Ascorbic acid sensor based on ion-sensitive field-effect transistor modified with MnO_2 nanoparticles. *Anal. Chim. Acta* **512**, 57–61. DOI: 10.1016/j.aca.2004.02.039

Luo X.-L., Xu J.-J., Zhao W., and Chen H.-Y. (2004) Glucose biosensor based on ENFET doped with SiO_2 nanoparticles. *Sens. Actuators B* **97**, 249–255. DOI: 10.1016/j.snb.2003.08.024

Luo X., Morrin A., Killard A.J., and Smyth M.R. (2006) Application of nanoparticles in electrochemical sensors and biosensors. *Electroanalysis* **18**(4), 319–326. DOI: 10.1002/elan.200503415

Massobrio G., Massobrio A., Massobrio L., and Massobrio P. (2011) Silicon-based biosensor functionalised with carbon nanotubes to investigate neuronal electrical activity in pH-stimulated environment: A modeling approach. *MicroNano Lett.* **6**(8), 8, 689–693. DOI: 10.1049/mnl.2011.0336

Massobrio G., Massobrio P., and Martinoia S. (2008) Modeling the neuron-carbon nanotube-ISFET junction to investigate the electrophysiological neuronal activity. *Nano Lett.* **8**(12), 4433–4440. DOI: 10.1021/nl802341r

Nair P.R. and. Alam M.A. (2006) Performance limits of nanobiosensors. *Appl. Phys. Lett.* **88**, 233120. DOI: 10.1063/1.2211310

Nair P.R., and. Alam M.A. (2007) Design considerations of silicon nanowire biosensors. *IEEE Trans. Electron. Dev.* **54**(12), 3400–3408. DOI: 10.1109/TED.2007.909059

Nair P.R. and Alam M.A. (2008) Screening-limited response of nano biosensors. *Nano Lett.* **8**(5), 1281–1285. DOI: 10.1021/nl072593i

Reddy B. Jr., Dorvel B.R., Go J., Nair P.R., Elibol O.H., Credo G.M., Chow E.K.C., Su X., Varma M., Alam M.A., and Bashir R. (2011) High-k dielectric Al_2O_3 nanowire and nanoplate field-effect sensors for improved pH sensing. *Biomed. Microdev.* **13**(2), 335–344. DOI: 10.1007/s10544-010-9497-z

Sarkar D. and Banerjee K. (2012) Proposal for tunnel-field-effect-transistor as ultra-sensitive and label-free biosensors. *Appl. Phys. Lett.* **100**, 143108-1 to 143108-4.

Schöning M.J. and Poghossian A. (2006) Bio FEDs (Field-effect devices): State-of-the-art and new directions. *Electroanalysis* **18**(19–20), 1893–1900. DOI: 10.1002/elan.200603609

Schöning M.J. and Poghossian A. (2008) Silicon-based field-effect devices for (bio-) chemical sensing. In: *Proceedings of ASDAM 2008, The Seventh International Conference on Advanced Semiconductor Devices and Microsystems*, October 12–16, 2008, Smolenice Castle, Slovakia, 31–38.

Shipway A.N. and Willner I. (2001) Nanoparticles as structural and functional units in surface-confined architectures. *Chem. Commun.* **2001**, 2035–2045. DOI: 10.1039/b105164b

Sone H., Fujinuma Y., and Hosaka S. (2004) Picogram mass sensor using resonance frequency shift of cantilever. *Jpn. J. Appl. Phys.* **43**(6A), 3648–3651.

Spijkman M., Smits E.C.P., Cillessen J.F.M., Biscarini F., Blom P.W.M., and de Leeuw D.M. (2011) Beyond the Nernst-limit with dual-gate ZnO ion-sensitive field-effect transistors. *Appl. Phys. Lett.* **98**, 043502. DOI: 10.1063/1.3546169

Stern E. (2007) Label-Free Sensing with Semiconducting Nanowires. Ph.D. dissertation, Yale University, New Haven, CT.

Stern E., Wagner R., Sigworth F.J., Breaker R., Fahmy T.M., and Reed M.A. (2007) Importance of the Debye screening length on nanowire field-effect transistor sensors. *Nano Lett.* **7**(11), 3405–3409. DOI: 10.1021/nl071792z

Ushaa S.M. and Eswaran V. (2012) Modeling and simulation of nanosensor arrays for automated disease detection and drug delivery unit. *Int. J. Adv. Eng. Technol.* **2**(1), 564–577.

Vieira N.C.S, Avansi W. Jr., Figueiredo A., Ribeiro C., Mastelaro V.R., and Guimarães F.E.G. (2012) Ion-sensing properties of 1D vanadium pentoxide nanostructures. *Nanoscale Res. Lett.* **7**, 310. DOI: 10.1186/1556-276X-7-310

Vijayalakshmi A., Tarunashree Y., Baruwati B., Manorama S.V., Narayana B.L., Johnson R.E.C., and Rao N.M. (2008) Enzyme field effect transistor (ENFET) for estimation of triglycerides using magnetic nanoparticles. *Biosens. Bioelectron.* **23**, 1708–1714. DOI: 10.1016/j.bios.2008.02.003

Wang Y. (2010) Simulation of a silicon nanowire FET biosensor for detecting biotin/streptavidin binding. M.S. thesis, University of Pittsburgh, Pittsburgh, PA.

Wang Y. and Li G. (2010) Performance investigation for a silicon nanowire FET biosensor using numerical simulation. In: *Proceedings of 2010 IEEE Nanotechnology Materials and Devices Conference,* Oct 12–15, Monterey, CA, 81–86.

Xu J.-J., Luo X.-L., and Chen H.-Y. (2005) Analytical aspects of FET-based biosensors. *Frontiers Biosci.* **10**, 420–430.

Xu J.-J., Zhao W., Luo X.-L., and Chen H.-Y. (2005) A sensitive biosensor for lactate based on layer-by-layer assembling of MnO_2 nanoparticles and lactate oxidase on ion-sensitive field-effect transistors. *Chem. Commun.* **2005**, 792–794. DOI: 10.1039/b416548a

Yao K., Zhu Y., Yang X., and Li C. (2008) ENFET glucose biosensor produced with dendrimer encapsulated Pt nanoparticles. *Mater. Sci. Eng.* C **28**, 1236–1241. DOI: 10.1016/j.msec.2007.11.004

Yates D.E., Levine S., and Healy T.W. (1974) Site-binding model of the electrical double layer at the oxide/water interface. *J. Chem. Soc., Faraday Trans.* **70**, 1807–1818, DOI: 10.1039/F19747001807

Zehfroosh N., Shahmohammadi M., and Mohajerzadeh S. (2010) High-sensitivity ion-selective field-effect transistors using nanoporous silicon. *IEEE Electron. Dev. Lett.* **31**(9), 1056–1058. DOI: 10.1109/LED.2010.2052344

Zhou S. and Wu H. (2012) Analytical solutions of nonlinear Poisson–Boltzmann equation for colloidal particles immersed in a general electrolyte solution by homotopy perturbation technique. *Colloid. Polym. Sci.* **290**, 1165–1180. DOI: 10.1007/s00396-012-2622-1

CHAPTER 9

Biosensors: Modeling and Simulation of Diffusion-Limited Processes

L. Rajendran

1. INTRODUCTION

1.1. ENZYME KINETICS

Kinetics is the study of rates of chemical reactions. Enzymes are little molecular machines that carry out reactions in cells. Enzymes are catalysts which reduce the needed activation energy so these reactions proceed at rates that are useful to the cell. Enzymes are catalysts (generally proteins) that help to convert other molecules called substrates into products, but they themselves are not changed by the reaction. Their most important features are catalytic power, specificity, and regulation. Enzymes accelerate the conversion of substrates into products by lowering the free energy of activation of the reaction. Enzyme kinetics is the study of rates of chemical reactions that involve enzymes. Living systems depend on ch.1emical reactions which, on their own, would occur at extremely slow rates. Study of enzyme kinetics (Cornish-Bowden 2004; Cook and Cleland 2007) is useful for measuring the concentration of an enzyme in a mixture (by its catalytic activity), its purity (specific activity), its catalytic efficiency and/or specificity for different substrates by comparison of different forms of the same enzyme in different tissues or organisms, and effects of inhibitors (which can give information about catalytic mechanism, structure of the active site, potential therapeutic agents, etc.).

DOI: 10.5643/9781606505984/ch9

1.2. BASIC SCHEME OF BIOSENSORS

According to IUPAC nomenclature, a biosensor (Figure 9.1) is a device that uses specific biochemical reactions mediated by isolated enzymes, immunosystems, tissues, organelles, or whole cells to detect chemical compounds, usually by electrical, thermal, or optical signals (McNaught and Wilkinson 1997). The beginning of biosensors may be dated to 1962, when Clark, known as the father of the biosensor concept, published an experiment in which glucose oxidase (GOx) was entrapped at a Clark oxygen electrode using a dialysis membrane (Clark and Lyons 1962).

As biocomponent, an enzyme, antibody, nucleic acid, lectine, hormone, cell structure, or tissue can be used. Its role is to interact specifically with the target analyte, and the result of biochemical reaction is subsequently transformed through a transducer to a measurable signal. The transducing system can be electrochemical, optical, piezoelectric, thermometric, ion-sensitive, magnetic, or acoustic. A very important part of biosensor fabrication is the immobilization of the biocomponent. The performance of biosensors (Kissinger 2005; Monosik et al. 2012) with immobilized molecules depends also on factors such as the chemical and physical conditions (pH, temperature, and contaminants), and the thickness and stability of the materials.

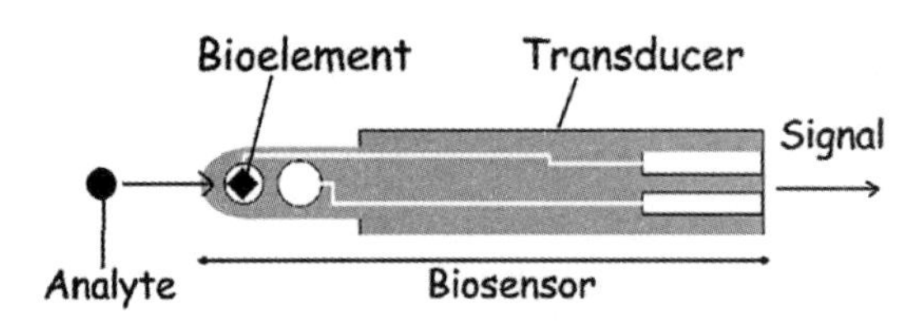

Figure 9.1. Basic scheme of a biosensor.

1.3. THE NONLINEAR REACTION-DIFFUSION EQUATION AND BIOSENSORS

Reaction-diffusion equations, semilinear diffusion equations, and free-boundary problems form an important domain of the theory of partial differential equations that is both very rich and challenging mathematically and is intricately related to numerous applications in physical, chemical, and biological sciences. Reaction-diffusion (RD) systems arise in many branches of physics, chemistry, biology, ecology, etc. Reviews of the theory and applications of reaction-diffusion systems can be found in books and numerous articles (see, e.g., (Cornish-Bowden 2004; Kissinger 2005; Cook and Cleland 2007; Monosik et al. 2012). These arise in a large variety of application areas, such as flow in porous media, heat conduction in plasmas, combustion problems, liquid evaporation, and, of more recent interest, image processing. Great effort is being put into development of the mathematical theory of nonlinear diffusion equations and to obtain exact solutions for special cases. Their significance relies not only on the huge number of their applications but also on the fact that they provide a rather general class of linear and nonlinear differential operators whose mathematical analysis has been shown to be a milestone for the development of applied, abstract, and numerical analysis, algebra, geometry, and topology.

1.3.1. Nonlinear Phenomena

Nonlinear phenomena play a crucial role in applied mathematics and chemistry. Exact (closed-form) solution of nonlinear reaction-diffusion equations plays an important role in the proper understanding of qualitative features of many phenomena and processes in various areas of natural science. The main result obtained from reaction and diffusion systems is that nonlinear phenomena include diversity of stationary and spatio-temporal dissipative patterns, oscillations, different types of waves, excitability, bistability, etc. However, it is difficult for us to obtain exact solutions for these problems. Investigation of exact solutions of nonlinear equations is both interesting and important. In general, this results in the need to solve linear and nonlinear reaction-diffusion equations with complex boundary conditions. The enzyme kinetics in biochemical systems have usually been modeled by differential equations which are based only on reaction, without spatial dependence of the various concentrations. The dimensionless nonlinear reaction diffusion equations are

$$\frac{\partial S}{\partial \tau} = \nabla^2 S - f(R,\tau,S,P)$$
$$\frac{\partial P}{\partial \tau} = \nabla^2 P + g(R,\tau,S,P) \tag{9.1}$$

where S and P represent the dimensionless concentrations of the active species, τ represents the dimensionless time, and R is the dimensionless radial coordinate of the particle. The first term on the right-hand side of Eq. (9.1) accounts for active species (substrate) diffusion, whereas the second term, $f(R, \tau, S, P)$ or $g(R, \tau, S, P)$, represents the homogeneous reaction term (nonlinear term), generally polynomial in the concentrations and time. In the following sections of this chapter, the behavior of the system is considered with both Michaelis-Menten and non–Michaelis-Menten kinetics. The biosensor action includes a heterogeneous enzymatic process and diffusion. The mathematical model involves a system of nonlinear differential equations with the inclusion of the enzymatic reaction and the diffusion of glucose (substrate) and acceptor (product).

1.4. TYPES OF BIOSENSORS

Biosensors are classified according to different criteria such as bioreceptors and transducers, and according to the various types of physical and chemical interaction. Depending on the type of transducer, biosensors can be classified as (1) electrochemical; (2) potentiometric/Piezometric; (3) thermal; or (4) optical. Examples of biosensor classification are shown in Figures 9.2 and 9.3.

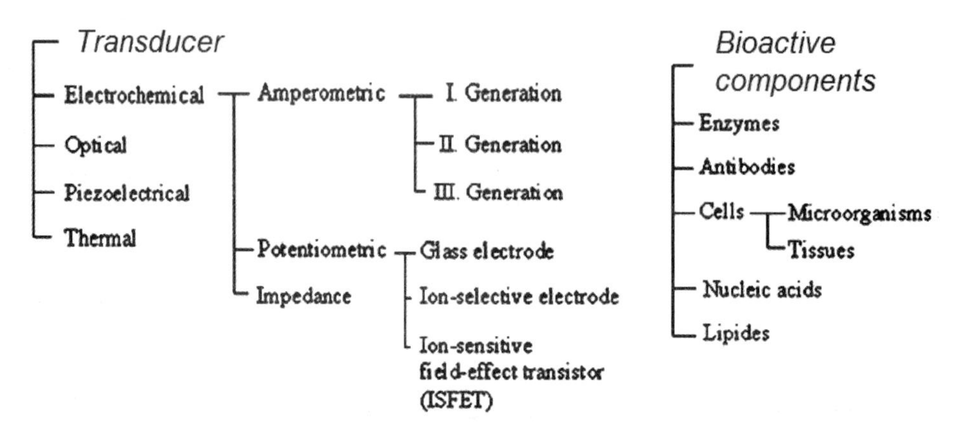

Figure 9.2. Different types of biosensors.

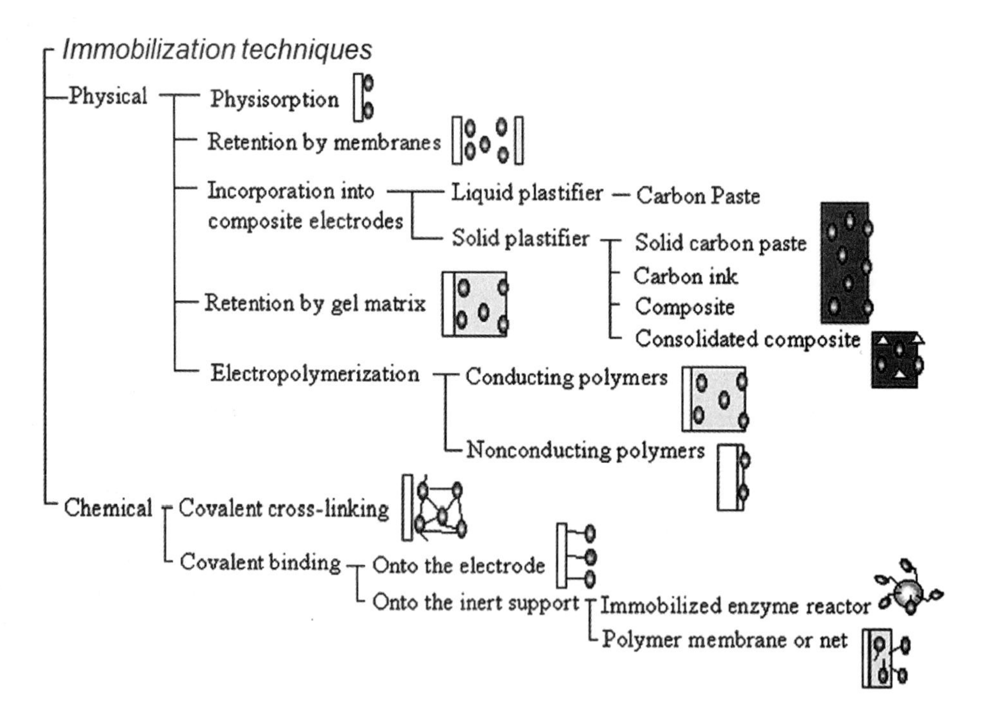

Figure 9.3. Different types of biosensors depending on interaction.

1.5. MICHAELIS-MENTEN KINETICS

In biochemistry, Michaelis-Menten kinetics is one of the simplest and best-known models of enzyme kinetics. It is named after German biochemist Leonor Michaelis and Canadian physician Maud Menten. The model takes the form of an equation describing the rate of enzymatic reactions, by relating reaction rate v to $[S]$ (the concentration of a substrate S). Its formula is given by

$$v = \frac{d[P]}{dt} = \frac{V_{\max}[S]}{k_m + [S]} \tag{9.2}$$

Here, $V_{\max}$ represents the maximum rate achieved by the system, at maximum (saturating) substrate concentrations. The Michaelis constant k_m is the substrate concentration at which the reaction rate is half of $V_{\max}$. Biochemical reactions involving a single substrate are often assumed to follow Michaelis-Menten kinetics, without regard to the model's underlying assumptions.

1.6. NON–MICHAELIS-MENTEN KINETICS

Some enzymes display non–Michaelis-Menten kinetics that do not approximate to Michaelis-Menten kinetics to any useful extent. In these cases there is little value in retaining the terminology and symbolism of Michaelis-Menten kinetics. Instead, it is often possible to express the rate as a rational function of the substrate concentration:

$$v = \frac{\alpha_1[A] + \alpha_2[A]^2 + \alpha_3[A]^3 + \cdots + \alpha_n[A]^n}{1 + \beta_1[A] + \beta_2[A]^2 + \beta_3[A]^3 + \cdots + \beta_m[A]^m} \tag{9.3}$$

(In principle, this kind of equation can be generalized to accommodate more than a single substrate, but it then becomes highly complicated and only the single-substrate case will be considered here.) A rational function is a ratio of two polynomials. The degree of a polynomial is the largest exponent; thus, the degree of the numerator of the expression in Eq. (9.3) is n and that of the denominator is m. The rational function as a whole may be described as an "n:m function." In general, the degree of the numerator of a rate equation does not exceed the degree of the denominator for enzyme-catalyzed reactions, but there is no other necessary relationship between n and m and neither bears any necessary relationship to the number of catalytic centers per molecule of enzyme. In the terminology of Eq. (9.3), any rate equation obeying Michaelis-Menten kinetics can be defined as a 1:1 function.

Under some conditions, which cannot be expressed simply and are not normally obvious from inspection of the coefficients of Eq. (9.3), the equation may generate a plot of v against $[A]$ that shows a monotonic increase in v toward a limiting value at all positive finite values of $[A]$. A necessary, but not sufficient, condition is that the degrees of the numerator and denominator be equal, i.e., $n = m$. It is then meaningful to define a limiting rate $V = \alpha_n/\beta_n$. Moreover, it may also be useful to describe the kinetics quantitatively in terms of the slope of a plot of $\log[v/(V - v)]$ against $\log[A]$, which is known as a Hill plot. This slope is called the Hill coefficient and is given the symbol h. In a kinetic context, it bears no necessary relationship of any kind to the number of catalytic centers per molecule of enzyme and it should not be given a symbol, such as n, that suggests that it does. At any concentration of substrate at which h is greater than unity, the kinetics are said to display *cooperativity*. In some contexts, the more explicit term *positive cooperativity* may be preferable to avoid ambiguity. At any concentration at which $h = 1$, the kinetics are said to be *noncooperative*, and if h is less than unity they are said to be *negatively cooperative*. In the case of Michaelis-Menten kinetics, $h = 1$ over the whole concentration range, but in other cases h is not constant and the sign of cooperativity may change one or more times over the range of concentrations. Thus cooperativity is not absolute and in general can only be defined in relation to a particular concentration.

1.7. IMPORTANCE OF MODELING AND SIMULATION OF BIOSENSORS

One of the main reasons restricting the wider use of biosensors is the relatively short linear range of the calibration curve (Nakamura and Karube 2003). Another serious drawback is the instability of biomolecules. These problems can be partially solved by the application of an additional outer perforated membrane (Turner et al. 1987; Scheller and Schubert 1992; Wollenberger et al. 1997). To improve the productivity and efficiency of a biosensor design as well as to optimize the biosensor configuration, a model of the real biosensor should be built (Amatore et al. 2006; Stamatin 2006). Modeling of a biosensor with a perforated membrane was performed by Schulmeister and Pfeiffer (1993). The proposed one-dimensional-in-space (1-D) mathematical model does not take into consideration the geometry of the membrane perforation, and it also includes effective diffusion coefficients. For a one-dimensional model, the quantitative value of diffusion coefficients is limited (Schulmeister and Pfeiffer 1993). Recently, a two-dimensional-in-space (2-D) mathematical model has been proposed that includes consideration of the perforation geometry (Baronas et al. 2006; Baronas 2007). However, simulation of the biosensor action based on the 2-D model is much more time-consuming than simulation based on the corresponding 1-D model. This is especially important when investigating numerically any peculiarities of the biosensor response over wide ranges of catalytic and geometric parameters. The multifold numerical simulation of the biosensor response based on the 1-D model is much more efficient than the simulation based on the corresponding 2-D model.

2. MODELING OF BIOSENSORS

2.1. MICHAELIS-MENTEN KINETICS AND POTENTIOMETRIC BIOSENSORS

Meena and Rajendran (2010) developed a mathematical model of steady-state and non–steady-state responses of a pH-based potentiometric biosensor immobilizing organophosphorus hydrolase (OPH). Figure 9.4 represents the scheme of the pH-based potentiometeric biosensor immobilizing OPH.

It is assumed that the substrate S reacts with the catalysts via Michaelis-Menten kinetics, according to the scheme $S + EH \underset{k_2}{\overset{k_1}{\Leftrightarrow}} EHS \xrightarrow{k_2} EHP \xrightarrow{k_3} E + P$. In this reaction scheme the nonlinear diffusion-reaction equation can be stated as

$$\frac{\partial S}{\partial t} = D\frac{\partial^2 S}{\partial x^2} \pm R(S) \tag{9.4}$$

$$R(S) = \frac{k_3[E]_t[S]}{\left(1+\frac{k_3}{k_2}\right)\left([S]+\frac{(k_2+k_{-1})\,k_3}{(k_2+k_3)\,k_1}\right)} \tag{9.5}$$

where S is the concentration of the species, D is the diffusion coefficient, and R is the reaction rate (nonlinear term). Using the following dimensionless variables,

$$\overline{[S]} = \frac{S}{k_m} \qquad \bar{x} = \frac{x}{L} \qquad a^2 = \frac{L^2 V_{\max}}{Dk_m} \qquad R = \frac{v_{\max}[S]}{[S]+k_m} \tag{9.6}$$

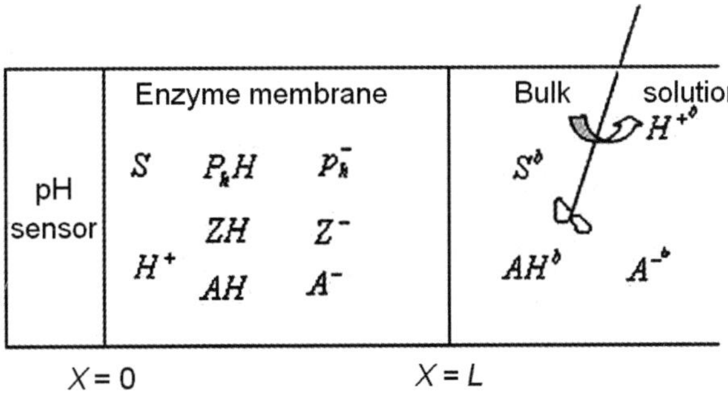

Figure 9.4. Scheme of the pH-based potentiometric biosensor immobilizing OPH. (Reprinted with permission from Meena and Rajendran 2010. Copyright 2010 WIley-VCH Verlag.)

The above equation (9.4) for steady state (Meena and Rajendran 2010) can be written as

$$\frac{d^2[\overline{S}]}{d\overline{x}^2} - a^2 \frac{[\overline{S}]}{[\overline{S}]+1} = 0 \tag{9.7}$$

The initial and boundary conditions may be presented as follows

$$[\overline{S}]\,(\overline{x},0) = 0 \qquad \left[\frac{\partial[\overline{S}]}{\partial\overline{x}}\right]_{\overline{x}=0} = 0 \tag{9.8}$$

The solution of the above equation for steady state is

$$[\overline{S}] = \frac{[\overline{S}]^b \cosh(a\overline{x})}{\cosh a} - \frac{([\overline{S}]^b)^2}{(\cosh a)^2}\left[1 - \frac{\cosh(2a\overline{x})}{3}\right]\left[\cosh(a\overline{x}) - \cosh a\right] \tag{9.9}$$

The above equation represents the simple and closed-form analytical expression of substrate concentration profiles for all values of the parameters.

2.2. MICHAELIS-MENTEN KINETICS AND AMPEROMETRIC BIOSENSORS

Rahamathunissa and Rajendran (2008) developed an application of He's variational iteration method in nonlinear boundary-value problems in enzyme–substrate reaction-diffusion processes in amperometric biosensors. In an irreversible mono-enzyme system, the reaction scheme for free enzyme E and substrate concentration S may be expressed by

$$E + S \overset{K_M}{\leftrightarrow} ES \xrightarrow{k_2} E + P \tag{9.10}$$

The one-dimensional form of this equation is

$$\frac{\partial S}{\partial t} = D_S \frac{\partial^2 S}{\partial \chi^2} - \frac{k_2 E_0 S}{K_M + S} \tag{9.11}$$

Introducing a pseudo-first-order rate constant $K = k_2E_0/K_M$, we can write Eq. (9.11) as

$$\frac{\partial S}{\partial t} = D_S \frac{\partial^2 S}{\partial \chi^2} - \frac{KS}{1 + S/K_M} \tag{9.12}$$

Here we consider an initial condition is given in the usual form, $S(0, \chi) = S_0$ $(\chi \in \Omega)$. The nonlinear partial differential equation (PDE) is made dimensionless by defining the following dimensionless parameters:

$$u = \frac{s}{K_S^\infty} \qquad x = \frac{\chi}{L} \qquad \tau = \frac{D_S t}{L^2} \qquad K = \frac{kL^2}{D_S} = \Phi^2 \qquad \alpha = \frac{kS^\infty}{K_M} \tag{9.13}$$

The steady-state dimensionless form of Eq. (9.12) is then

$$\frac{\partial^2 u}{\partial x^2} - \frac{Ku}{1+\alpha u} = 0 \qquad 0 \le u \le 1 \tag{9.14}$$

The initial condition reduces to $u(x = 0) = a$ (constant). The boundary condition is

$$u(0) = a = \begin{cases} \sec h(\sqrt{k}) & \text{for } \alpha << 1 \\ 1 - k/2\alpha & \text{for } \alpha >> 1 \end{cases} \tag{9.15}$$

Using the variational iteration method, the solution to the above problem is

$$u(x) = a + \frac{Kx^2}{2\alpha} + \frac{(1+\alpha a)}{a\alpha^2}\ln\left[\frac{Ka\alpha}{2(1+\alpha a)^2}x^2 + 1\right] - \frac{\sqrt{2K}}{\alpha^{\frac{3}{2}}\sqrt{a}}x\tan^{-1}\left[\frac{\sqrt{Ka\alpha}}{\sqrt{2}(1+\alpha a)}x\right] \tag{9.16}$$

Equation (9.16) represents the new analytical expression for substrate concentration profiles for all values of α and K. The concentration $u(x)$ versus distance is plotted in Figure 9.5 for various values of parameters.

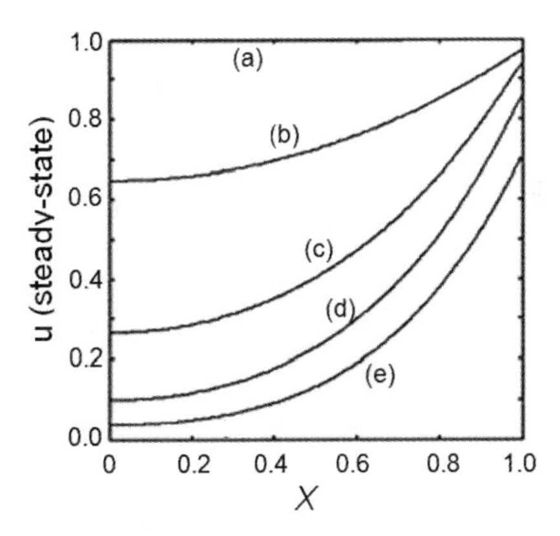

Figure 9.5. Plot of normalized steady-state substrate concentration u versus normalized distance x when $\alpha u < 1$ (for $\alpha = 0.1$) for various values of K: (a) $K = 0.01$; (b) $K = 1$; (c) $K = 4$; (d) $K = 9$; (e) $K = 16$.

2.3. MICHAELIS-MENTEN KINETICS AND AMPEROMETRIC BIOSENSORS FOR IMMOBILIZING ENZYMES

Anitha et al. (2011b) developed analytical solutions of amperometric enzymatic reactions. Figure 9.6 represents a general kinetic scheme of possible mechanisms occurring at an enzyme electrode. The differential equations that quantify the diffusion and reaction within the film may be written as

$$\frac{de_2}{dt} = \frac{k_{\text{cat}} s(e_\Sigma - e_Z)}{K_M + s} - k_a e_Z$$

$$D_S \frac{d^2 s}{dx^2} - \frac{k_{\text{cat}} s(e_\Sigma - e_Z)}{K_M + s} = 0$$

$$D_B \frac{d^2 b}{dx^2} + kae_Z = 0 \tag{9.17}$$

where D_S is the diffusion coefficient of the substrate into the polymer film, D_B is the diffusion coefficient of the oxidized mediator into the polymer film, K_M is the Michaelis-Menten constant, k_{cat} is the catalytic reaction rate constant, k_a is the

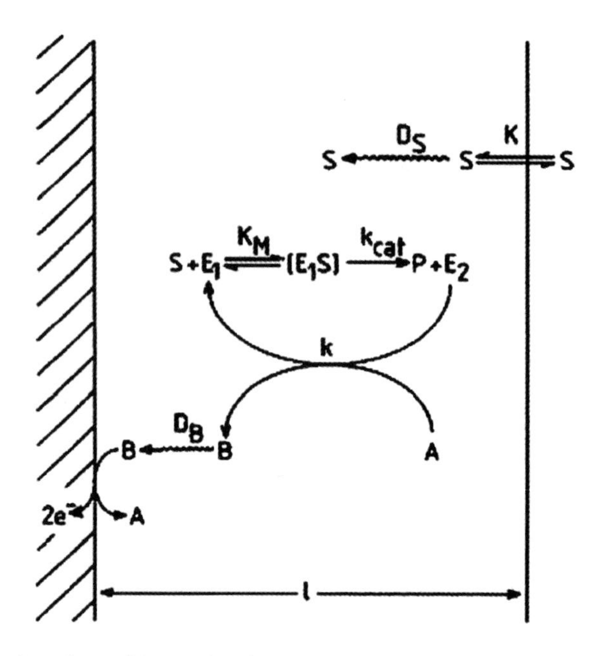

Figure 9.6. Schematics of possible mechanisms occurring at an enzyme electrode. (Reprinted with permission from Anitha et al. 2011. Copyright 2011 Elsevier; and Reprinted with permission from Bartlett and Whitaker 1987. Copyright 1987 Elsevier.)

regeneration of the oxidized enzyme, x is the distance from the electrode/polymer interface, and $e_\Sigma = e_1 + e_2$ is the total enzyme concentration in the film. Applying the steady-state condition, we obtain the following nonlinear reaction diffusion equations in dimensionless form:

$$\frac{d^2s}{dx^2} - \frac{ks}{1+\alpha s} = 0 \tag{9.18}$$

$$\frac{d^2b}{dx^2} - \frac{\eta s}{1+\alpha s} = 0 \tag{9.19}$$

where $\alpha = (k_{cat} + k_a)/K_M k_a$; $\kappa = k_{cat} e_\Sigma / D_S k_m$; $\eta = k_{cat} e_\Sigma / D_B k_m$. The boundary conditions are reduced such that at $x = 0$, $ds/dx = 0$, $b = 0$; and at $x = l$, $s = Ks_\infty$, $b = 0$. Using the homotopy perturbation method (HPM), the analytical expressions of the concentrations are

$$s(x) = \frac{(Ks_\infty)\cosh(\sqrt{kx})}{\cosh(\sqrt{kx})} + \frac{\alpha(Ks_\infty)^2\{\cosh(\sqrt{kl})[3-\cosh(2\sqrt{kx})] - \cosh(\sqrt{kx})[3-\cosh(2\sqrt{kl})]\}}{6\cosh(\sqrt{kx})^3} \tag{9.20}$$

$$\begin{aligned} b(x) = {} & \frac{\eta(Ks_\infty)[(1-\cosh(\sqrt{kx}) - (1-\cosh(\sqrt{kl})x/l]}{k\cosh(\sqrt{kl})} + \frac{\alpha\eta(Ks_\infty)^2[\cosh(\sqrt{kx}) - \cosh(\sqrt{kl})x/l]}{2k\cosh(\sqrt{kl})^3} \\ & + \left[\frac{\alpha\eta(Ks_\infty)^2\cosh(2\sqrt{kl})}{6k\cosh(\sqrt{kl})^3} - \frac{\alpha\eta(Ks_\infty)^2}{6k\cosh(\sqrt{kl})^2} - \frac{\alpha\eta(Ks_\infty)^2}{2k\cosh(\sqrt{kl})^3}\right]\left(1-\frac{x}{l}\right) \\ & - \frac{\alpha\eta(Ks_\infty)^2[\cosh(2\sqrt{kl})\cosh(\sqrt{kx}) - \cosh(\sqrt{kl})\cosh(2\sqrt{kx})]}{6k\cosh(\sqrt{kl})^3} \end{aligned} \tag{9.21}$$

Equations (9.20) and (9.21) represent the analytical expressions of the substrate and mediator concentrations for all values of parameters.

2.4. MICHAELIS-MENTEN KINETICS AND THE TWO-SUBSTRATE MODEL

Indira and Rajendran (2011) developed an analytical expression of the concentration of substrates and product in a phenol–polyphenol oxidase (PPO) system immobilized in laponite hydrogels. Figure 9.7 shows a schematic representation of a rotating disk electrode modified by a PPO enzymatic layer of thickness L.

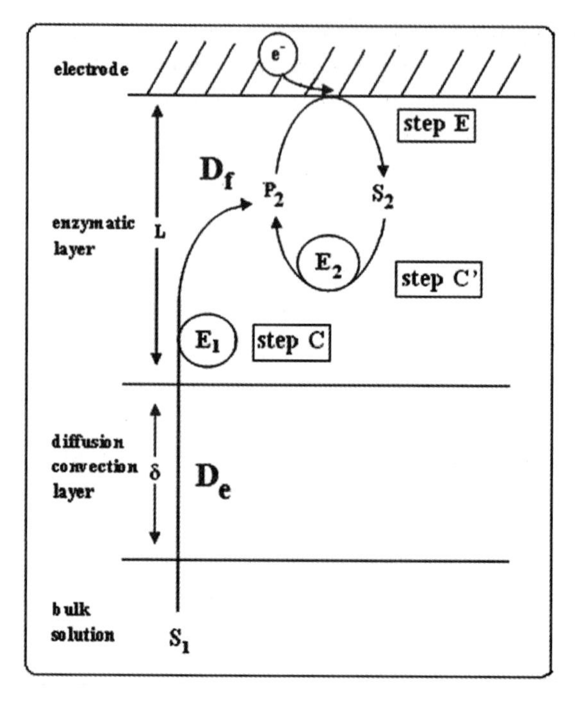

Figure 9.7. Schematic representation of rotating-disk electrode modified by a PPO enzymatic layer and principle of the bioelectrode functioning in the presence of phenol substrates S_1, S_2 and product P_2. (Reprinted with permission from Indira and Rajendran 2011. Copyright 2011 Elsevier.)

The different steps which lead to the electrode response are as follows. (1) Here we can assume that the diffusion coefficients of the phenol substrate S_1, the catechol substrate S_2, and the *o*-quinone product P_2 are equal in the bulk solution, due to their structural similarity. (2) Diffusion of S_1, S_2, and P_2 within the enzyme layer of thickness L with a similar diffusion coefficient D_f and with the same partition coefficient κ. (3) The enzymatic reaction of substrates S_1, S_2, and the product P_2 follows Michaelis-Menten formalism in a homogeneous medium, which is $V_1 = k_1[E_T]\,[S_1]/K_1 + [S_1]$ and $V_2 = k_2[E_T]\,[S_2]/K_2 + [S_2]$. Here V_1 and V_2 represent the enzymatic rates of *o*-quinone formation from phenol and catechol substrates, respectively. $[E_T]$ represents the total concentration of active enzyme. K_1, K_2, k_1, and k_2 are the kinetic parameters. The electroenzymatic process can be written as follows:

$$\text{Step C:}\qquad S_1 + O_2 \xrightarrow{\text{PPO}} P_2 + H_2O \qquad V_1$$

$$\text{Step E:}\qquad P_2 + 2e^- + 2H^+ \underset{K_O}{\overset{K_r}{\rightleftarrows}} S_2 \qquad E^0$$

$$\text{Step C}':\qquad S_2 + \frac{1}{2}O_2 \xrightarrow{\text{PPO}} P_2 + H_2O \qquad V_2$$

The nonlinear differential equations for Michaelis-Menten formalism described by the substrates and product concentrations within the enzymatic layer at steady state are

$$\frac{\partial^2[S_1]}{\partial X^2} - \frac{[S_1]}{\Lambda_1^{\ 2}(1+[S_1]/K_1)} = 0$$
$$\frac{\partial^2[S_2]}{\partial X^2} - \frac{[S_2]}{\Lambda_2^{\ 2}(1+[S_2]/K_2)} = 0$$
$$\frac{\partial^2[P_2]}{\partial X^2} + \frac{[S_1]}{\Lambda_1^{\ 2}(1+[S_1]/K_1)} + \frac{[S_2]}{\Lambda_2^{\ 2}(1+[S_2]/K_2)} = 0 \quad (9.22)$$

where X represents the distance from the electrode surface, and Λ_1 and Λ_2 are the reaction lengths related to S_1 and S_2. Here $\Lambda_1 = (D_f\, K_1/k_1[E_T]^{1/2})$, $\Lambda_2 = (D_f\, K_2/k_2[E_T]^{1/2})$, where D_f is the diffusion coefficient in the enzymatic layer. The Λ_i describe the distance over which the substrate S_i can diffuse in the enzyme layer undergoing enzymatic reaction. These equations are solved for the following boundary conditions: $[S_1] + [S_2] = [S_1]_\infty$, $[P_2] = 0$, $X = 0$ and $[S_1] = [S_1]_\infty$, $[S_2] = 0$, $[P_2] = 0$, $X = L + \delta$. In addition, mass conservation implies that when $X \geq L^+$, $[S_1] + [S_2] + [P_2] = [S_1]_\infty$, and when $X \leq L^-$, $[S_1] + [S_2] + [P_2] = \kappa[S_1]_\infty$, where L is the thickness of the enzymatic layer and $\delta\ (= 1.61D_e^{\ 1/3}\upsilon^{1/6}\omega^{-1/2})$ is the thickness of the diffusion convection layer and κ is the partition coefficient for the primary reactant in the rotating-disk electrode system. At $X = L$, the fluxes J_{S_1}, J_{S_2}, and J_{P_1} of S_1, S_2, and P_2 on each side of the interface are equal.

We introduce the set of dimensionless variables $x = X/L$, $u = [S_1]/[S_1]_\infty$, $v = [S_2]/[S_1]_\infty$, $w = [P_2]/[S_1]_\infty$, where u, v, and w represent dimensionless concentrations and x is the distance parameter. We get the dimensionless nonlinear equations for the rotating-disk electrode as follows:

$$\frac{\partial^2 u}{\partial x^2} - \frac{\mu_1 u}{1+\alpha_1 u} = 0$$
$$\frac{\partial^2 v}{\partial x^2} - \frac{\mu_2 v}{1+\alpha_2 v} = 0$$
$$\frac{\partial^2 w}{\partial x^2} + \frac{\mu_1 u}{1+\alpha_1 u} + \frac{\mu_2 v}{1+\alpha_2 v} = 0 \quad (9.23)$$

where $\mu_1 = L^2/\Lambda_1^{\ 2}$, $\mu_2 = L^2/\Lambda_2^{\ 2}$, $\alpha_1 = [S_1]_\infty/K_1$, and $\alpha_2 = [S_1]_\infty/K_2$. Here, μ_1 is the ratio of the square of the thickness L and the reaction length Λ_1, μ_2 is the ratio of the square of the thickness L and the reaction length Λ_2, α_1 is the ratio of $[S_1]_\infty$ and the kinetic parameter K_1, α_2 is the ratio of $[S_1]_\infty$ and the kinetic parameter K_2. Now the initial and boundary conditions are represented as follows. The dimensionless current is given by

$$I = \frac{j_f}{2F} = -\left(\frac{\partial v}{\partial x}\right)_{x=0} \tag{9.24}$$

where F is the Faraday constant. Using the homotopy perturbation method, the solutions of Eqs. (9.23) are

$$u(x) = \frac{\cosh\left(\sqrt{\mu_1}x\right)}{\cosh\left(\sqrt{\mu_1}m\right)} + \frac{\alpha_1}{2\cosh^2\left(\sqrt{\mu_1}m\right)}\left\{\frac{\cosh\left(\sqrt{\mu_1}x\right)}{\cosh\left(\sqrt{\mu_1}m\right)}\left[\frac{\cosh\left(2\sqrt{\mu_1}m\right)}{3} - 1\right] - \frac{\cosh\left(2\sqrt{\mu_1}x\right)}{3} + 1\right\} \tag{9.25}$$

$$v(x) = \frac{\sinh\left[\sqrt{\mu_2}(m-x)\right]}{\sinh\left(\sqrt{\mu_2}m\right)}\left\{1 - \frac{1}{\cosh\left(\sqrt{\mu_1}m\right)} - \frac{\alpha_1}{6\cosh^2\left(\sqrt{\mu_1}m\right)}\left[\frac{\cosh\left(2\sqrt{\mu_1}m\right)-3}{\cosh\left(\sqrt{\mu_1}m\right)} + 2\right]\right\}$$

$$+ \frac{\alpha_2}{\sinh^2\left(\sqrt{\mu_2}m\right)}\left[1 - \frac{1}{\cosh\left(\sqrt{\mu_1}m\right)}\right]$$

$$\times\left\{\frac{\sinh\left[\sqrt{\mu_2}(m-x)\right]}{2\sinh\left(\sqrt{\mu_2}m\right)}\left[\frac{\cosh\left(2\sqrt{\mu_2}m\right)}{3} + 1\right] + \frac{2\sinh\left(\sqrt{\mu_2}x\right)}{3\sinh\left(\sqrt{\mu_2}m\right)} - \frac{\cosh\left[2\sqrt{\mu_2}(m-x)\right]}{6} - \frac{1}{2}\right\}^2 \tag{9.26}$$

$$w(x) = 1 - \frac{\cosh\left(\sqrt{\mu_1}x\right)}{\cosh\left(\sqrt{\mu_1}m\right)} - \frac{\sinh\left[\sqrt{\mu_2}(m-x)\right]\left\{1-\left[1/\cosh\left(\sqrt{\mu_1}m\right)\right]\right\}}{\sinh\left(\sqrt{\mu_2}m\right)}$$

$$-\frac{(\mu_1\alpha_2 + \mu_2\alpha_1)\left\{1-\left[1/\cosh\left(\sqrt{\mu_1}m\right)\right]\right\}}{2\cosh\left(\sqrt{\mu_1}m\right)(\mu_1-\mu_2)^2}$$

$$\left[\left(\frac{\left(\sqrt{\mu_1}+\sqrt{\mu_2}\right)^2\sinh\left[\left(\sqrt{\mu_1}-\sqrt{\mu_2}\right)x + \sqrt{\mu_2}m\right] - \left(\sqrt{\mu_1}-\sqrt{\mu_2}\right)^2\sinh\left[\left(\sqrt{\mu_1}+\sqrt{\mu_2}\right)x - \sqrt{\mu_2}m\right]}{\sinh\left(\sqrt{\mu_2}m\right)}\right)\right.$$

$$\left. + \frac{2x}{m}\left[(\mu_1+\mu_2) - \frac{2\sqrt{\mu_1}\sqrt{\mu_2}\sinh\left(\sqrt{\mu_1}m\right)}{\sinh\left(\sqrt{\mu_2}m\right)}\right] - 2(\mu_1+\mu_2)\right] \tag{9.27}$$

Equations (9.25)–(9.27) represent the dimensionless concentration of the substrates and product in the phenol–polyphenol oxidase system. When α_1 and $\alpha_2 \to 0$ (or K_1 and $K_2 \to \infty$), the concentrations u, v, and w become

$$u(x)=\frac{\cosh\left(\sqrt{\mu_1}x\right)}{\cosh\left(\sqrt{\mu_1}m\right)}\qquad v(x)=\left[1-\frac{1}{\cosh\left(\sqrt{\mu_1}m\right)}\right]\frac{\sinh\left[\sqrt{\mu_2}(m-x)\right]}{\sinh\left(\sqrt{\mu_2}m\right)}$$

and

$$w(x)=1-\frac{\cosh\left(\sqrt{\mu_1}x\right)}{\cosh\left(\sqrt{\mu_1}m\right)}-\left[1-\frac{1}{\cosh\left(\sqrt{\mu_1}m\right)}\right]\frac{\sinh\left[\sqrt{\mu_2}(m-x)\right]}{\sinh\left(\sqrt{\mu_2}m\right)}\tag{9.28}$$

Using Eq. (9.24), we can obtain the current as follows:

$$\begin{aligned}I=&\frac{\sqrt{\mu_2}\alpha_2}{\sinh^2\left(\sqrt{\mu_2}m\right)}\left[\frac{1}{\cosh\left(\sqrt{\mu_1}m\right)}-1\right]^2\\&\left\{\frac{2}{3\sinh\left(\sqrt{\mu_2}m\right)}+\frac{\sinh\left(2\sqrt{\mu_2}m\right)}{3}-\frac{\cosh\left(\sqrt{\mu_2}m\right)}{2\sinh\left(\sqrt{\mu_2}m\right)}\left[\frac{\cosh\left(2\sqrt{\mu_2}m\right)}{3}+1\right]\right\}\\&+\frac{\sqrt{\mu_2}\cosh\left(\sqrt{\mu_2}m\right)}{\sinh\left(\sqrt{\mu_2}m\right)}\left\{1-\frac{1}{\cosh\left(\sqrt{\mu_1}m\right)}-\frac{\alpha_1}{6\cosh^2\left(\sqrt{\mu_1}m\right)}\left[\frac{\cosh\left(2\sqrt{\mu_1}m\right)-3}{\cosh\left(\sqrt{\mu_1}m\right)}+2\right]\right\}\end{aligned}\tag{9.29}$$

Figure 9.8 shows the series of normalized concentration profiles for all values of α_1 and δ/L. From Figure 9.8*a*, it is evident that the value of the phenol substrate concentration u decreases when the thickness of the enzymatic layer increases and enzyme activity increases for all values of α_1 and m. Figure 9.8*b* shows the normalized concentration profile of the catechol substrate v. It is clear that the catechol substrate concentration v increases when the thickness of the enzymatic layer increases. Figure 9.8*c* represents the normalized *o*-quinone product concentration w. It is clear that the concentration w increases when the thickness of the enzymatic layer increases.

2.5. NON–MICHAELIS-MENTEN KINETICS

2.5.1. Amperometric Biosensor with Mixed Enzyme Kinetics

Manimozhi et al (2010) presents the solution of steady-state substrate concentration in the action of biosensor response with mixed enzyme kinetics under a

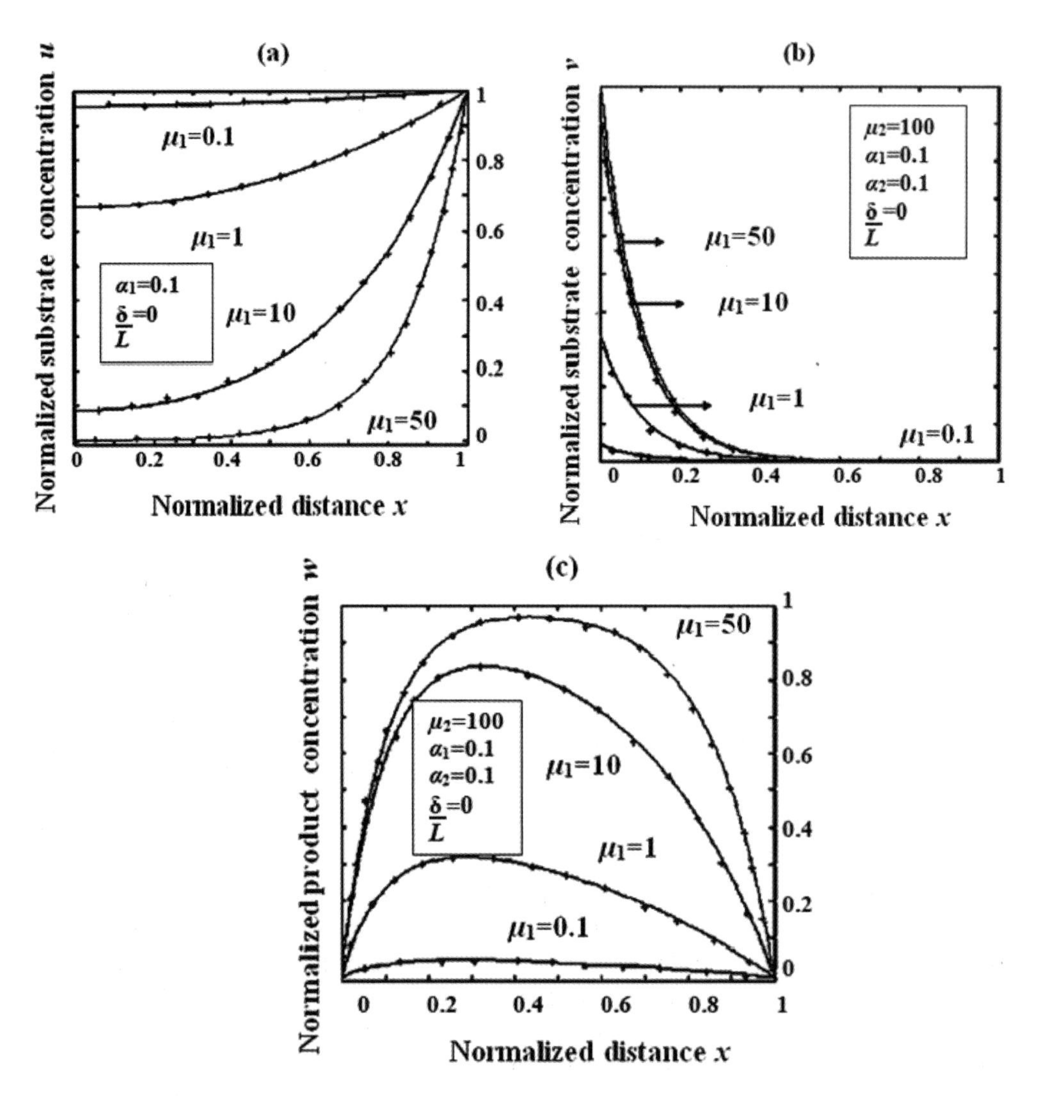

Figure 9.8. Plot of normalized concentration profiles versus normalized distance. Symbols: (−) analytical solution and (+) numerical simulation. From this figure, it is evident that the value of concentration u decreases and v, w increases when the thickness of the enzymatic layer increases and enzyme activity increases (μ_1, μ_2 increases) for all values of α_1 and δ/L.

Michaelis-Menten scheme for the interaction of the enzyme–substrate complex [ES] with other substrate molecules [S] following the generation of nonactive complex [ES2], as

$$\mathrm{E} + \mathrm{S} \underset{k_2}{\overset{k_1}{\longleftrightarrow}} \mathrm{ES} \xrightarrow{k_c} \mathrm{E} + \mathrm{P} \qquad \text{and} \qquad \mathrm{E} + \mathrm{S} \longleftrightarrow \mathrm{ES}_2$$

Based on the non–Michaelis-Menten hypothesis, the velocity function v for the simple reaction process without competitive inhibition is given by

$$v = \frac{V_{\max}[S]}{K_M + [S] + [S]^2 / K_i}$$

where $V_{\max} = k_c[E]_0$; K_M and K_i are the Michaelis-Menten and inhibition constants, respectively. In this model, the equation for S becomes

$$\frac{\partial S}{\partial t} = D_S \frac{\partial^2 S}{\partial \chi^2} - \frac{K_C E_0 S}{K_M + S + S^2 / K_i} \tag{9.30}$$

Introducing a pseudo-first-order rate constant $K = K_C E_0 / K_M$, the above equation becomes

$$\frac{\partial S}{\partial t} = D_S \frac{\partial^2 S}{\partial \chi^2} - \frac{K_C E_0 S}{1 + S / K_M + S^2 / K_i K_M} \tag{9.31}$$

The equation must be solved subject to the initial and boundary conditions: $t = 0$, $S = 0$; $\chi = 0$, $\partial S/\partial \chi = 0$; $\chi = 1$, $S = 1$. The system governs the substrate concentration S when there is no competitive inhibition in the reaction. The nonlinear PDE is made dimensionless by defining the following parameters:

$$u = \frac{S}{ks^{\infty}} \qquad x = \frac{\chi}{L} \qquad \tau = \frac{D_s t}{L^2} \qquad k = \frac{KL^2}{D_s} = \varphi^2 \qquad \alpha = \frac{ks^{\infty}}{K_M} \qquad \beta = \frac{ks^{\infty}}{K_i K_M} \tag{9.32}$$

where u, x, and t represent dimensionless concentration, distance, and time, respectively. Here α denotes a saturation parameter and K denotes reaction diffusion parameter, and the above equation reduces to the following dimensionless form:

$$\frac{\partial u}{\partial t} = \frac{\partial^2 u}{\partial x^2} - \frac{Ku}{1 + \alpha u + \beta u^2} \qquad 0 < u \le 1 \tag{9.33}$$

For the case of steady state,

$$\frac{\partial^2 u}{\partial x^2} - \frac{Ku}{1 + \alpha u + \beta u^2} = 0 \tag{9.34}$$

The boundary conditions reduces to $u = 1$, $x = 1$, and $\partial u/\partial x = 0$, $x = 0$. The solution of the equation thus becomes

$$u(x) = u_1(x) = 1 - a + Ax^2 + Bx^4 + Cx^6 \tag{9.35}$$

where

$$A = \frac{k(1-a)}{2} - a[\alpha(1-a) + \beta(1-a)^2]$$

$$B = \frac{ak - 2a^2[\alpha + 2\beta(1-a)]}{12}$$

$$C = -\frac{\beta a^3}{15} \tag{9.36}$$

where a is the solution of the equation

$$4.4\beta a^3 - 5\begin{pmatrix}\alpha + \\ 2\beta\end{pmatrix} a^2 + \begin{pmatrix}6a + 6\beta + \\ 2.5K + 6\end{pmatrix} a - 3K = 0$$

Equation (9.35) is the simple approximate analytical expression corresponding to the concentration of substrate for all values of reaction/diffusion parameters.

2.6. OTHER ENZYME REACTION MECHANISMS

2.6.1. Simple Enzyme Kinetics Reaction Mechanisms

Meena et al (2010) developed mathematical modeling of enzyme kinetics reaction mechanisms and derived the analytical solutions of nonlinear reaction equations. The reactant molecule that binds to the enzyme is termed the substrate S, and the mechanism is often written as

$$\mathrm{E} + \mathrm{S} \underset{k-1}{\overset{k_1}{\longleftrightarrow}} \mathrm{ES} \xrightarrow{k_2} \mathrm{E} + \mathrm{P}$$

The concentrations of the reactants are denoted by lowercase letters, $s = [S]$, $e = [E]$, $c = [SE]$, $p = [P]$. The nonlinear reaction equations are given by

$$\frac{ds}{dt} = -k_1 es + k_{-1}c \qquad \frac{de}{dt} = -k_1 es + (k_{-1} + k_2)c$$

$$\frac{dc}{dt} = k_1 es - (k_{-1} + k_2)c \qquad \frac{dp}{dt} = k_2 c \tag{9.37}$$

The boundary conditions are $s(0) = s0$, $e(0) = e0$, $c(0) = 0$, $p(0) = 0$. In the above equation, $de/dt + dc/dt = 0$ and $e(t) + c(t) = e_0$. The system of ordinary differential equations reduces to only two, for s and c:

$$\frac{ds}{dt} = -k_1 e_0 s + (k_{1s+}k_{-1})c \qquad \frac{dc}{dt} = k_1 e_0 s - (k_{-1} + k_1 s + k_2) \tag{9.38}$$

with initial conditions $s(0) = s0$, $c(0) = 0$. Introducing the following parameters,

$$\tau_1 = \frac{k_1 e_0 t}{\varepsilon} \qquad u(\tau) = \frac{s(t)}{s_0} \qquad v(\tau) = \frac{c(t)}{e_0} \qquad w(\tau) = \frac{p(t)}{e_0},$$
$$\lambda = \frac{k_2}{k_1 s_0} \qquad k = \frac{k_{-1} + k_2}{k_1 s_0} \qquad \varepsilon = \frac{e_0}{s_0} \tag{9.39}$$

the dimensionless form of the equation is

$$\frac{du}{d\tau} = -u\varepsilon + \varepsilon(u + k - \lambda)v \qquad \frac{dv}{d\tau} = u - (u + k)v \qquad \frac{dw}{d\tau} = \lambda v \tag{9.40}$$

with initial conditions $u(0) = 1$, $v(0) = 0$, $w(0) = 0$. Using the variational iteration method, the solutions of the above equations are

$$u(\tau) = e^{-\varepsilon\tau} - \frac{e^{-\varepsilon\tau}}{k(k-\varepsilon)^2}\left\{\begin{matrix} e^{-k\tau}\varepsilon^2 - k\varepsilon\left[\varepsilon(\lambda - k)\tau + e^{-\varepsilon\tau} + e^{-k\tau}\right] - \varepsilon(\lambda - k)k \\ +k\left[ke^{-\varepsilon\tau} + \varepsilon(\lambda - k)k\tau + \varepsilon(\lambda - k)e^{(\varepsilon - k)\tau}\right] - (\varepsilon - k)^2 \end{matrix}\right\}$$
$$v(\tau) = \left(\frac{e^{-\varepsilon\tau} - e^{-k\tau}}{k - \varepsilon}\right) - \left(\frac{e^{-k\tau}}{k - \varepsilon}\right)\left[\frac{(e^{-k\tau} - 1)}{\varepsilon} + \frac{(e^{(k-2\varepsilon)} - 1)}{k - 2\varepsilon}\right] \tag{9.41}$$

Equation (9.41) represents the analytical expressions of the concentration of substrate $u(\tau)$ and enzyme–substrate complex $v(\tau)$ for all values of time.

2.6.2. Second-Order Reaction Mechanism

Varadharajan and Rajendran (2011a) developed an analytical solution of coupled nonlinear second-order reaction differential equations in enzyme kinetics. The reaction

$$\mathrm{E} + \mathrm{S} \underset{k_{-1}}{\overset{K_1}{\Leftrightarrow}} \mathrm{C} \xrightarrow{K_2} \mathrm{E} + \mathrm{P}$$

is the binding of substrate S and release of product P. The nonlinear differential equations are

$$\frac{d[S]}{dt} = k_1\left[-([E_0]-[C][S]) + K_S[C]\right]$$

$$\frac{d[C]}{dt} = k_1\left[([E_0]-[C][S]) + K_M[C]\right]$$

$$\frac{d[P]}{dt} = k_2[C] \tag{9.42}$$

and by imposing the laws of mass action, $[E_0] = [E](t) + [C](t)$, $[S_0] = [S](t) + [C](t) + [P](t)$, with initial conditions at $t = 0$, $[S] = [S_0]$, $[E] = [E_0]$, $[C] = 0$, $[P] = 0$. In this system the parameters k_1, k_{-1}, and k_2 are positive rate constants, $K_S = k_{-1}/k_1$ is the equilibrium dissociation constant, $K = k_2/k_1$ is the Van Slyke-Cullen constant, and $K_M = K_S + K$ is known as the Michaelis-Menten constant. The dimensionless parameters are

$$\tau = \frac{k_1[E_0](t)}{\varepsilon} \qquad u(\tau) = \frac{[S](t)}{[S_0]} \qquad v(\tau) = \frac{[C](t)}{[E_0]} \qquad w(\tau) = \frac{[P](t)}{[E_0]}$$

$$\lambda = \frac{k_2}{k_1[S_0]} \qquad k = \frac{K_M}{[S_0]} \qquad \varepsilon = \frac{[E_0]}{[S_0]} \tag{9.43}$$

The dimensionless equations are

$$\frac{du}{d\tau} = u\varepsilon + \varepsilon(u + k - \lambda)v \qquad \frac{dv}{d\tau} = u - (u + k)v \qquad \frac{dw}{d\tau} = \lambda v \tag{9.44}$$

and the initial conditions are $u(0) = 1,\ v(0) = 0,\ w(0) = 0$. The solutions are

$$u(\tau) = e^{-\varepsilon\tau} + \frac{e^{-\varepsilon\tau}}{k} + \left(\frac{\varepsilon}{k-\varepsilon}\right)\left(\frac{e^{-(k+\varepsilon)\tau}}{k} - \frac{e^{-2\varepsilon\tau}}{\varepsilon}\right) + \left[\frac{\varepsilon(k-\lambda)\tau e^{-\varepsilon\tau}}{k-\varepsilon}\right] + \frac{\varepsilon(k-\lambda)\left(e^{-k\tau} - e^{-\varepsilon\tau}\right)}{(k-\varepsilon)^2}$$

$$v(\tau) = \left(\frac{e^{-\varepsilon\tau} - e^{-k\tau}}{k-\varepsilon}\right) + \frac{e^{-k\tau}}{\varepsilon(-2\varepsilon)} - \left[\frac{e^{-2\varepsilon\tau}}{(k-\varepsilon)(k-2\varepsilon)}\right] - \frac{e^{-(k+\varepsilon)\tau}}{\varepsilon(k-\varepsilon)} \tag{9.45}$$

$$w(\tau) = \frac{\lambda\left(1-e^{-\varepsilon\tau}\right)}{\varepsilon(k-\varepsilon)} - \frac{\lambda\left(1-e^{-k\tau}\right)}{\varepsilon(k-\varepsilon)} + \frac{\lambda\left(1-e^{-\varepsilon\tau}\right)}{k\varepsilon(k-2\varepsilon)} - \frac{\lambda\left(1-e^{-2\varepsilon\tau}\right)}{2\varepsilon(k-\varepsilon)(k-2\varepsilon)} - \frac{\lambda\left(1-e^{-(k+\varepsilon)\tau}\right)}{\varepsilon(k+\varepsilon)(k-\varepsilon)}$$

Equation (9.45) represents the analytical expression of the dimensionless substrate concentration u, enzyme substrate concentration v, and product concentration w for all values of parameters.

2.6.3. Enzyme–Substrate Reaction Diffusion Processes

Meena et al. (2011) developed a mathematical modeling of biosensors with enzyme–substrate interaction and biomolecular interaction. The reaction is

$$E + S \underset{k_{-1}}{\overset{K_1}{\Leftrightarrow}} C \xrightarrow{K_{cat}} E + P$$

The system of nonlinear reaction equations is

$$D_s \frac{d^2 s}{dx^2} - k_1 es + k_{-1}c = 0$$

$$D_e \frac{d^2 e}{dx^2} - k_1 es + (k_{-1} + k_{cat})c = 0$$

$$D_c \frac{d^2 c}{dx^2} + k_1 es - (k_{-1} + k_{cat})c = 0$$

$$D_p \frac{d^2 p}{dx^2} + k_{cat}c = 0 \tag{9.46}$$

where k_1 is the forward rate of complex formation and k_{-1} is the backward rate constant. The boundary conditions when $t > 0$ are $x = 0$, $ds/dx = 0$, $dp/dx = 0$, $de/dx = 0$, $dc/dx = 0$; when $t > 0$ they are $x = L$, $s = s_0$, $dp/dx = 0$, $de/dx = 0$, $dc/dx = 0$, and adding, $d^2e/dx^2 + d^2c/dx^2 = 0$. Using the boundary conditions and from the law of mass conservation, we obtain $e = e_0 - c$. The system of ordinary differential equations reduces to only two equations, for s and c, that is,

$$D \frac{d^2 s}{dx^2} - k_1 e_0 s + (k_1 s + k_{-1})c = 0$$

$$D \frac{d^2 c}{dx^2} + k_1 e_0 s - (k_1 s + k_{-1} + k_{cat})c = 0 \tag{9.47}$$

Now, introducing dimensionless parameters,

$$u = \frac{s}{s_0} \qquad v = \frac{c}{e_0} \qquad w = \frac{p}{e_0} \qquad X = \frac{x}{L}$$

$$\gamma_S = \frac{k_{-1}L^2}{D} \qquad \gamma_E = \frac{k_1 s_0 L^2}{D} \qquad \gamma_P = \frac{k_{cat}L^2}{D} \tag{9.48}$$

The dimensionless forms of the differential equation are as follows:

$$\frac{d^2u}{dX^2} - \gamma_E u + (\gamma_S + \gamma_E u)v = 0$$

$$\frac{d^2v}{dX^2} + \gamma_E u - (\gamma_S + \gamma_E u + \gamma_P)v = 0$$

$$\frac{d^2w}{dX^2} + \gamma_P v = 0 \tag{9.49}$$

where γ_E, γ_S, and γ_P are the dimensionless reaction diffusion parameters. The boundary conditions when $X = 0$ are $du/dX = 0$, $dv/dX = 0$, $dw/dX = 0$; and when $X = 1$ they are $u = 1$, $dv/dX = 0$, $dw/dX = 0$. Using the variational iteration method, the concentrations of the substrate and the enzyme–substrate are

$$\begin{aligned} u(X) = {} & 1 - a + 0.5\gamma_E\left[1 + ab - b - (\gamma_S/\gamma_E)b - a\right]X^2 + 0.083\gamma_E\left[2a - 1 - ab - (\gamma_S/\gamma_E)\right]X^4 \\ & + 0.1\gamma_E\left[1 - a + (\gamma_S/\gamma_E)\right]X^5 - 0.03\gamma_E\left[1 + (\gamma_S/\gamma_E)\right]X^6 + 0.05a\gamma_E X^7 - 0.02a\gamma_E X^8 \end{aligned}$$

$$\begin{aligned} v(X) = {} & 0.02a\gamma_E X^8 - 0.05a\gamma_E X^7 + 0.03\gamma_E\left\{1 + \left[(\gamma_S/\gamma_E) + (\gamma_P/\gamma_E)\right]\right\}X^6 \\ & + 0.1\gamma_E\left\{a - 1 - \left[(\gamma_S/\gamma_E) + (\gamma_P/\gamma_E)\right]\right\}X^5 + 0.083\gamma_E\left\{\begin{matrix}\left[(\gamma_S/\gamma_E) + (\gamma_P/\gamma_E)\right] \\ -2a + 1 + ab\end{matrix}\right\}X^4 \\ & + 0.5\gamma_E\left\{b - ab + \left[(\gamma_S/\gamma_E) + (\gamma_P/\gamma_E)\right]b - 1 + a\right\}X^2 + b \end{aligned} \tag{9.50}$$

where

$$a = \frac{7}{5}\left[\frac{\gamma_S + \gamma_P - 29\gamma_E + 30b(\gamma_S + \gamma_P + \gamma_E)}{\gamma_E(28b - 27)}\right]$$

and

$$b = 0.5\left[(2100(\gamma_S + \gamma_E) + 10500\gamma_P)\gamma_E\right]^{-1}$$
$$\left[\begin{matrix} 9820\gamma_E\gamma_P - 25200\gamma_S + 2000\gamma_S\gamma_E + 4100\gamma_E^2 + 25200\gamma_P \\ -20\left(\begin{matrix} 1595160\gamma_E\gamma_S\gamma_P + 3175200(\gamma_E\gamma_S + \gamma_E\gamma_P + \gamma_P\gamma_S) + 1587600(\gamma_E^2 + \gamma_P^2 + \gamma_S^2) + 3175200\gamma_P\gamma_S \\ +112564\gamma_E^2\gamma_P\gamma_S - 786240\gamma_E^2\gamma_P + 1325520\gamma_E\gamma_P^2 + 276676\gamma_E^2\gamma_P^2 - 1676\gamma_E^3\gamma_P + 269640\gamma_E\gamma_S^2 \\ +274680\gamma_S\gamma_E^2 + 11449\gamma_E^2\gamma_S^2 + 428\gamma_E^3\gamma_S + 5040\gamma_E^3 + 4\gamma_E^4 \end{matrix}\right)^{1/2} \end{matrix}\right]$$

The dimensionless concentration of enzyme is

$$e(X) = e(X)/e_0 = 1 - v(X) = 1 - b + 0.5\gamma_E\{b - ab + [(\gamma_S/\gamma_E) + (\gamma_P/\gamma_E)]b - 1 + a\}X^2$$
$$+ 0.083\gamma_E\{[(\gamma_S/\gamma_E) + (\gamma_P/\gamma_E)] - 2a + 1 + ab\}X^4 + 0.1\gamma_E\{a - 1 - [(\gamma_S/\gamma_E) + (\gamma_P/\gamma_E)]\}X^5$$
$$+ 0.03\gamma_E\{1 + [(\gamma_S/\gamma_E) + (\gamma_P/\gamma_E)]\}X^6 - 0.05a\gamma_E X^7 + 0.02a\gamma_E X^8 \tag{9.51}$$

The dimensionless concentration of the product is given by

$$w(X) = 0.0335X^2 + 0.0167\gamma_P X^2 - 0.0333X^2(X-1)^2 - 0.0667X^3 - 0.0833\gamma_P X^4$$
$$+ 0.0333X^4 + 0.1000\gamma_P X^5 - 0.00033\gamma_P X^6 \tag{9.52}$$

Equation (9.50) represents the analytical expressions of the substrate $u(X)$ and enzyme–substrate $v(X)$ concentration for all values of parameters.

2.7. KINETICS OF ENZYME ACTION

Varadharajan and Rajendran (2011b) developed an analytical solution of a system of nonlinear differential equations for a single-enzyme, single-substrate reaction with non–mechanism-based enzyme inactivation. The single enzyme–substrate reaction system is

$$S + E \underset{k_{-1}}{\overset{K_1}{\Leftrightarrow}} C \xrightarrow{K_{cat}} P + E \qquad \text{with} \qquad E \xrightarrow{k_3} E_i$$

The nonlinear differential equations are

$$\frac{ds}{dt} = -k_1 s(e_0 - c - e_i) + k_1 K_S c$$
$$\frac{dc}{dt} = k_1 s(e_0 - c - e_i) - k_1 K_M c$$
$$\frac{de_i}{dt} = k_1 K_\delta (e_0 - c - e_i) \tag{9.53}$$

with initial conditions at $t = 0$ of $s = s_0$, $c = 0$, $e_i = 0$. The dimensionless parameters are

$$\tau = \frac{k_1 e_0 t}{\varepsilon} \qquad u(\tau) = \frac{s(t)}{s_0} \qquad v(\tau) = \frac{c(t)}{e_0} \qquad w(\tau) = \frac{e_i(t)}{e_0}$$
$$\lambda_1 = \frac{K_S}{s_0} \qquad \lambda_2 = \frac{K_M \varepsilon}{e_0} \qquad \lambda_3 = \frac{K_\delta \varepsilon}{e_0} \qquad \varepsilon = \frac{e_0}{s_0} \tag{9.54}$$

In dimensionless form, Eqs. (9.53) become

$$\frac{du}{d\tau} = -u\varepsilon + \varepsilon uv + \varepsilon uw + \lambda_1 \varepsilon v$$

$$\frac{dv}{d\tau} = u - uv - uw + \lambda_2 v$$

$$\frac{dw}{d\tau} = \lambda_3 - \lambda_3 v - \lambda_3 w \tag{9.55}$$

The boundary conditions are $u(0) = 1$, $v(0) = 0$, $w(0) = 0$. The solutions of the equations are

$$u(\tau) = e^{-\varepsilon\tau} + \left(\frac{\lambda_1 \varepsilon \tau e^{-\varepsilon\tau}}{\lambda_2 - \varepsilon}\right) + \frac{\lambda_1 \varepsilon \left(e^{-\lambda_2 \tau} - e^{-\varepsilon\tau}\right)}{(\lambda_2 - \varepsilon)^2} + \varepsilon\tau e^{-\varepsilon\tau} + \frac{\varepsilon\left[e^{-(\lambda_3+\varepsilon)\tau} - e^{-\varepsilon\tau}\right]}{\lambda_3}$$

$$v(\tau) = \frac{e^{-\lambda_2 \tau}}{\varepsilon(\lambda_2 - \varepsilon)} - \frac{e^{-(\lambda_2+\varepsilon)\tau}}{\varepsilon(\lambda_2 - \varepsilon)} + \frac{\left(e^{-\lambda_2 \tau} - e^{-2\varepsilon\tau}\right)}{(\lambda_2 - \varepsilon)\,(\lambda_2 - 2\varepsilon)} + \frac{\left[e^{-(\lambda_3+\varepsilon)\tau} - e^{-\lambda_2 \tau}\right]}{\lambda_2 - \lambda_3 - \varepsilon}$$

$$w(\tau) = 1 - e^{-\lambda_3 \tau} + \frac{\lambda_3\left(e^{-\lambda_3 \tau} - e^{-\varepsilon\tau}\right)}{(\lambda_2 - \varepsilon)\,(\lambda_3 - \varepsilon)} + \frac{\lambda_3 e^{-\lambda_2 \tau}}{(\lambda_2 - \varepsilon)\,(\lambda_3 - \lambda_2)} - \frac{e^{-\lambda_3 \tau}}{(\lambda_2 - \varepsilon)\,(\lambda_3 - \lambda_2)} \tag{9.56}$$

Equation (9.56) represents the analytical expressions of the dimensionless concentrations of substrate, enzyme–substrate, and free enzyme for all values of parameters.

2.8. TRIENZYME BIOSENSOR

A general scheme that represents the reaction occurring at a creatinine biosensor is shown as

$$S \xrightarrow{E_1} P_1 \xrightarrow{E_2} P_2 \xrightarrow{E_3} P_3$$

Here s_1, p_1, and p_2 represent the concentrations of S, P_1, and P_2, which are less than Michaelis-Menten constants ($K_{m(i)}$). The diffusion equations with a constant diffusion coefficient corresponding to the enzymatic conversion at a non–steady-state condition have the following forms (Anitha et al. 2011a):

$$\frac{1}{D}\frac{\partial s_1}{\partial t} = \frac{\partial^2 s_1}{\partial x^2} - \alpha_1^2 s_1 \tag{9.57}$$

$$\frac{1}{D}\frac{\partial p_1}{\partial t} = \frac{\partial^2 p_1}{\partial x^2} + \alpha_1^2 s_1 - \alpha_2^2 p_1 \tag{9.58}$$

$$\frac{1}{D}\frac{\partial p_2}{\partial t} = \frac{\partial^2 p_2}{\partial x^2} + \alpha_2^2 p_1 - \alpha_3^2 p_2 \tag{9.59}$$

$$\frac{1}{D}\frac{\partial p_3}{\partial t} = \frac{\partial^2 p_3}{\partial x^2} + \alpha_3^2 p_2 \tag{9.60}$$

where D is the diffusion coefficient of all compounds; t is the time and $K_{m(i)}$ denote Michaelis-Menten constants; $\alpha_i = \left(V_{\max(i)}/K_{m(i)}D\right)^{1/2}$, where i = 1, 2, and 3, and $V_{\max(i)}$ denote the maximal enzymatic rates. The initial and boundary conditions are as follows: when $t = 0$, $s_1 = s_0$, $p_1 = 0$, $p_2 = 0$, $p_3 = 0$; when $x = d$, $s_1 = s_0$, $p_1 = 0$, $p_2 = 0$, $p_3 = 0$; and when $x = 0$, $\partial s_1/\partial x = 0$, $\partial p_1/\partial x = 0$, $\partial p_2/\partial x = 0$, $p_3 = 0$, where d is the membrane thickness. The biosensor current i is defined as

$$i = 2FD\left(\frac{\partial p_3}{\partial x}\right)_{x=0} \tag{9.61}$$

where F is the Faraday number. The analytical solution of Eq. (9.57) is

$$s_1(x,t) = \frac{s_0\cosh(\alpha_1 x)}{\cosh(\alpha_1 d)} + \frac{4s_0\alpha_1^{\,2}}{\pi}\sum_{n=0}^{\infty}\frac{\cos(lx)\exp(-(\alpha_1^{\,2}+l^2)Dt)}{(-1)^n(2n+1)(\alpha_1^{\,2}+l^2)} \tag{9.62}$$

The analytical solutions of Eqs. (9.58)–(9.60) are reported and discussed by Anitha et al. (2011). When time tends to infinity, the analytical expression of non–steady-state concentration [Eq. (9.62)] approaches the steady-state value, thereby confirming the validity of the mathematical analysis.

3. MICRODISK BIOSENSORS

3.1. INTRODUCTION

Amperometric biosensors (Turner 1987; Scheller and Schubert 1992) measure the current that arises on a working electrode by direct electrochemical oxidation or reduction of the biochemical reaction product. The current is proportionate to the concentration of the target analyte. However, amperometric biosensors possess a number of serious drawbacks.

One of the main reasons that restricts the wider use of amperometric biosensors is the relatively short linear range of the calibration curve (Nakamura and Karube 2003). Another serious drawback is the instability of biomolecules. These problems can be partially solved by the application of an additional outer perforated membrane (Turner 1987; Scheller and Schubert 1992; Nakamura and Karube 2003). To improve the productivity and efficiency of a biosensor design as well as to optimize the biosensor configuration, a model of the real biosensor should be built (Amatore et al. 2006; Stamatin et al. 2006). Modeling of a biosensor with a perforated membrane has been performed by Schulmeister and Pfeiffer (1993). The proposed one-dimensional model does not take into consideration the geometry of the membrane. For a one-dimensional model, the quantitative value of diffusion coefficients is limited (Schulmeister and Pfeiffer 1993). Recently, a two-dimensional mathematical model has been proposed that takes into consideration the perforation geometry (Baronas 2006; Baronas 2007).

Analytical solutions for the steady-state current at a microdisk chemical sensor have been reported by Dong and Che (1991) and by Lyons et al. (1998). Galceran and co-workers (2001) have described the current at a microdisk where an enzyme is present in bulk solution, but a model for immobilized enzymes on microdisks has not so far been reported. Phanthong and Somasundrum (2003) obtained an approximate expression of steady-state concentration and current in integral form for a microdisk biosensor when Michaelis-Menten constant K_M is large. Recently, Eswari and Rajendran (2010b) derived the concentration profile of the product of the enzyme reaction and the electrode current for all values of K_M using the homotopy perturbation method.

3.2. MATHEMATICAL FORMULATION OF THE PROBLEM

A polymer/enzyme solution is considered to be the case of a microdisk electrode in a modified form of drop coating. We assume the polymer/enzyme droplet takes a hemispherical shape on an insulating plane, into which the microdisk is inlaid. It is assumed that the film radius is larger than the radius of the disk, so the analysis is simplified by taking the microdisk to a microhemisphere as shown in Figure 9.9.

This is the same approximation as used by Dong and Che (1991) and by Phanthong and Somasundrum (2003) for a microdisk electrode modified by a redox polymer. It is assumed that the enzyme concentration is uniform and the enzyme reaction follows Michaelis-Menten kinetics, in which case the reaction in the film is (Phanthong and Somasundrum 2003)

$$S + E_1 \underset{k_2}{\overset{k_1}{\Leftrightarrow}} [E_1S] \xrightarrow{k_{cat}} P + E_2$$

If the solution is stirred uniformly, so that S is constantly supplied to the film, at steady state the mass balance for S will be given by

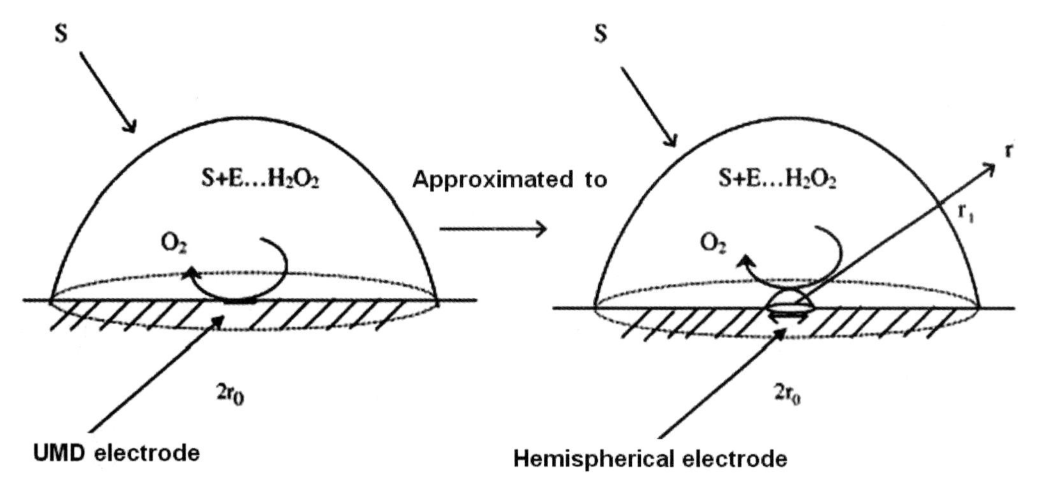

Figure 9.9. Coordinate system for microdisk electrode. E = enzyme, S = substrate. H_2O_2 is a reaction of the type given by Eq. (9.63). (Reprinted with permission from Eswari and Rajendran 2010. Copyright 2010 Elsevier.)

$$\frac{D_S}{r^2}\frac{d}{dr}\left(r^2\frac{dc_S}{dr}\right)-\frac{k_{\text{cat}}c_E c_S}{c_S+K_M}=0 \tag{9.63}$$

where c_S is the concentration profile of substrate, c_E is the concentration profile of enzyme, D_S is its diffusion coefficient, and K_M is the Michaelis constant, defined as $K_M = (k_{-1} + k_{\text{cat}})/k_1$. The steady-state mass balance for H will be given by

$$\frac{D_H}{r^2}\frac{d}{dr}\left(r^2\frac{dc_H}{dr}\right)+\frac{k_{\text{cat}}c_E c_S}{c_S+K_M}=0 \tag{9.64}$$

where c_H is the concentration profile of hydrogen peroxide. Then the boundary conditions at the electrode surface (r_0) and at the film surface (r_1) are given by

$$\begin{aligned} r = r_0: &\quad \frac{dc_S}{dr}=0 \qquad c_H = 0 \\ r = r_1: &\quad c_S = c_S^* \qquad c_H = 0 \end{aligned} \tag{9.65}$$

where c_S^* is the bulk concentration of S scaled by the partition coefficient of the film. The steady-state flux at the electrode surface is

$$\frac{I}{nFA}=D_H\left(\frac{dc_H}{dr}\right)_{r=r_0} \tag{9.66}$$

3.3. FIRST-ORDER CATALYTIC KINETICS

In this case, the substrate concentration c_S is less than the Michaelis constant K_M. Then Eqs. (9.63) and (9.64) reduce to

$$\frac{D_S}{r^2}\frac{d}{dr}\left(r^2\frac{dc_S}{dr}\right)-\frac{k_{\text{cat}}c_E c_S}{K_M}=0 \quad (9.67)$$

$$\frac{D_H}{r^2}\frac{d}{dr}\left(r^2\frac{dc_H}{dr}\right)+\frac{k_{\text{cat}}c_E c_S}{K_M}=0 \quad (9.68)$$

Solving Eq. (9.67) using the reduction of order (Ramana 2007), the concentration of substrate c_S can be obtained:

$$\frac{c_S}{c_S^*}=\frac{r_1[\alpha\exp(\chi r)+\exp(-\chi r)]}{r[\alpha\exp(\chi r_1)+\exp(-\chi r_1)]} \quad (9.69)$$

where

$$\alpha=\frac{(\chi r_0+1)\exp(-2\chi r_0)}{(\chi r_0-1)} \quad (9.70)$$

and

$$\chi=\sqrt{\frac{k_{\text{cat}}c_E}{D_S K_M}} \quad (9.71)$$

When χr_1 is small (χr is also small), $c_S/c_S^* = 1$. Similarly, when χr_1 is large [χr and $\chi(r_1 - r)$ are also large], $c_S/c_S^* = 0$. Using the reduction of order (Eswari and Rajendran 2010a), we obtain the concentration profile of hydrogen peroxide c_H as

$$\frac{D_H c_H}{D_S c_S^*}=\left\{\frac{r_1(r_1-r)}{r(r_1-r_0)}\left(\frac{\alpha e^{\chi r_0}+e^{-\chi r_0}}{\alpha e^{\chi r_1}+e^{-\chi r_1}}\right)+\left[\frac{r_1(r-r_0)}{r(r_1-r_0)}\right]-\left[\frac{r_1(\alpha e^{\chi r}+e^{-\chi r})}{r(\alpha e^{\chi r_1}+e^{-\chi r_1})}\right]\right\} \quad (9.72)$$

Equation (9.72) represents a general analytical approximation of the concentration of hydrogen peroxide, c_H, for all values of parameters χr_0 and r_1/r_0. We obtain the current as

$$\frac{Ir_0}{nFAD_S c_S^*}=g(\chi r_0, r_1/r_0)=\frac{r_1\left[\frac{(\alpha e^{\chi r_0}+e^{-\chi r_0})r_0-(\alpha e^{\chi r_1}+e^{-\chi r_1})r_0}{(r_0-r_1)}-\chi r_0\left(\alpha e^{\chi r_0}-e^{-\chi r_0}\right)\right]}{r_0(\alpha e^{\chi r_1}+e^{-\chi r_1})} \quad (9.73)$$

Equation (9.73) represents the new simple analytical expression of current for all values of χr_0 and r_1/r_0.

3.3.1. Comparison with the Work of Phanthong and Somasundrum (2003)

Recently, Phanthong and Somasundrum (2003) derived an analytical expression of steady-state concentration of hydrogen peroxide c_H [Eq. (9.74)] and steady-state current [Eq. (9.75)] when $c_S < K_M$ in integral form:

$$\frac{D_H c_H}{D_S c_S^*} = \frac{r_1}{[\alpha\exp(\chi r_1)+\exp(-\chi r_1)]}\left\{-\alpha\int_{r_0}^{r_1}\frac{(\chi r-1)\exp(\chi r)}{r^2}dr - \int_{r_0}^{r_1}\frac{(-\chi r-1)\exp(-\chi r)}{r^2}dr\right.$$

$$\left.+\left[\alpha\int_{r_0}^{r_1}\frac{(\chi r-1)\exp(\chi r)}{r^2}dr + \int_{r_0}^{r_1}\frac{(-\chi r-1)\exp(-\chi r)}{r^2}dr\right]\frac{r_1(r-r_0)}{r(r_1-r_0)}\right\} \tag{9.74}$$

$$\frac{Ir_0}{nFAc_S^*D_S} = g(\chi r_0, r_1/r_0) = \frac{r_1^2\left[\alpha\int_{r_0}^{r_1}\frac{(\chi r-1)\exp(\chi r)}{r^2}dr\int_{r_0}^{r_1}\frac{(-\chi r-1)\exp(-\chi r)}{r^2}dr+\right]}{(r_1-r_0)[\alpha\exp(\chi r_1)+\exp(-\chi r_1)]} \tag{9.75}$$

Phanthong and Somasundrum (2003) also obtained the empirical expression of current:

$$g(\chi r_0, r_1/r_0) = \frac{r_1}{r_1-r_0}\left\{1-\operatorname{sech}\left[(\chi r_0)^p\left(\frac{r_1-r_0}{r_1}\right)^q\right]\right\} \tag{9.76}$$

where p and q are empirical constants. The values of p and q are given for various values of χr_0 in Table 9.1. This empirical expression is also compared to our simple closed analytical expression [Eq. (9.73)] in Table 9.1. The average relative difference between our Eq. (9.73) and the empirical expression [Eq. (9.76)], is 1.7% when $\sigma = 1.1$. By analyzing the approximations of Eqs. (9.72) and (9.74), our approximation Eq. (9.72) is found to be the simplest form of concentration. Similarly, among the approximations of Eqs. (9.73) and (9.75), our approximation Eq. (9.73) is found to be the simplest closed form of current.

Table 9.1 represents the comparison of dimensionless current $g(x, \sigma)$ for various values of χr_0 for thin film ($\sigma = r_1/r_0 = 1.1$) using Eqs. (9.74) and (9.77).

The diffusion coefficient, also called diffusivity, is an important parameter indicative of the diffusion mobility. The diffusion coefficient is encountered not only

Table 9.1. Comparison of dimensionless current $g(x, \sigma)$ for various values of χr_0 for thin film ($\sigma = r_1/r_0 = 1.1$) using Eqs. (9.74) and (9.77)

χr_0	$\sigma = r_1/r_0$	p	q	α	Eq. (9.77) (Phanthong and Somasundrum 2003)	Eq. (9.74) (this work)	Error (%)
7	1.1	0.99	1	0.0000	2.1650	2.126	1.8350
5.5	1.1	0.985	1	0.0000	1.4113	1.3985	0.9157
5	1.1	0.983	1	0.0000	1.1847	1.1773	0.6264
4.5	1.1	0.983	1	0.00019	0.9795	0.9699	0.9947
4	1.1	0.983	1	0.00056	0.7892	0.7783	1.4026
3.5	1.1	0.978	1	0.00164	0.6082	0.6042	0.6627
3	1.1	0.96	1	0.00496	0.4383	0.4493	2.4512
2.5	1.1	0.97	1	0.0157	0.3175	0.3153	0.7074
2	1.1	0.955	1	0.0549	0.2035	0.2035	0.0035
1.5	1.1	0.92	1	0.2489	0.1150	0.1152	0.2028
0.9	1.1	1.5	0.99	−3.1407	0.0419	0.0417	0.3633
0.8	1.1	1.4	0.98	−1.8171	0.0322	0.0329	2.1056
0.7	1.1	1.35	0.97	−1.3974	0.0241	0.0253	4.8926
0.6	1.1	1.23	0.97	−1.2047	0.0179	0.0186	3.5037
0.5	1.1	1.17	0.97	−1.1036	0.0125	0.0129	3.4055
0.4	1.1	1.1	0.98	−1.0484	0.0080	0.0083	3.2720
0.3	1.1	1.1	0.97	−1.0192	0.0045	0.0046	2.9211
0.2	1.1	1.05	0.98	−1.0055	0.0021	0.0021	2.2225
0.1	1.1	1.04	0.98	−1.0007	0.0005	0.0005	0.3174
Average percent deviation							1.7266

in Fick's law, but also in numerous other equations of physics and chemistry. It is one of the most important characteristics of molecules in solutions. Diffusion coefficients in electrolyte solutions can be determined accurately by means of chronoamperometric methods with a thin-walled hanging-mercury-drop electrode or a stationary-disk electrode (Ikeuchi and Sato 1991). When the enzyme kinetic term is very small (χr_0 and $\chi r_1 \to 0$, i.e., it is purely diffusion-limited), Eq. (9.73) becomes

$$\frac{Ir_0}{nFAc_S^* D_S} = g(\chi r_0, r_1/r_0) = 2 \tag{9.77}$$

When the enzyme kinetic term is very large (χr_0 and $\chi r_1 \to \infty$), Eq. (9.73) becomes

$$\frac{Ir_0}{nFAc_S^* D_S} = g(\chi r_0, r_1/r_0) = \frac{r_1}{r_1 - r_0} \tag{9.78}$$

From the above equations, we can find the values of the diffusion coefficient. From the above two equations we can also obtain the current for all values of parameter χr_0:

$$\frac{Ir_0}{nFAc_S^* D_S} = \frac{2 + 2\chi r_0}{1 + 2\chi r_0 - (2r_0 \chi r_0 / r_1)} \tag{9.79}$$

3.3.2. Discussion when $c_S < K_M$

Figure 9.10 shows the dimensionless concentration profile of hydrogen peroxide, $D_H c_H / D_S c_S^*$, using Eq. (9.72) for all various values of χr_0 and r_1/r_0. Thus it is concluded that there is a simultaneous increase in the values of the concentration of hydrogen peroxide as well as in χr_0 and r_1/r_0. Figure 9.11 shows the dimensionless steady-state current $g(\chi r_0, r_1/r_0)$ using Eq. (9.73) for all values of χr_0. From this figure it is inferred that the value of the current increases when χr_0 increases.

The current attains the maximum value at $r_1/r_0 = 1.15$ (when $\chi r_0 = 20$), $r_1/r_0 = 1.18$ (when $\chi r_0 = 10$) and $r_1/r_0 = 1.25$ (when $\chi r_0 = 5$). Current is slowly decreasing

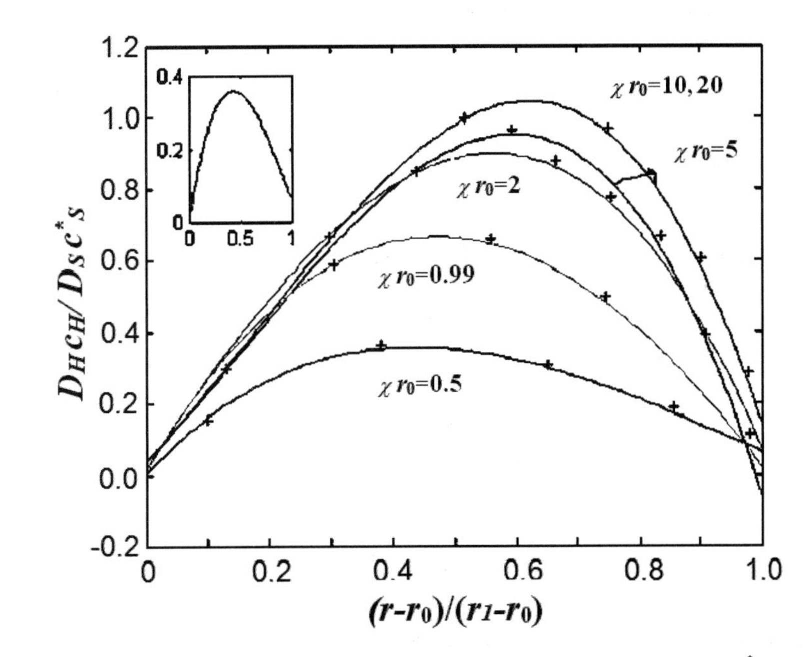

Figure 9.10. Dimensionless concentration profile of hydrogen peroxide, $D_H c_H / D_S c_S^*$, for different enzyme reaction/substrate diffusion ratios χr_0, when $r_1/r_0 = 1.5$. Key: (--) represents Eq. (9.73) and (+) represents Eq. (9.75) (Phanthong and Somasundrum 2003). Inset: Plot for $\chi r_0 = 0.5$ shown with expanded scale. Vertical line is midpoint of film.

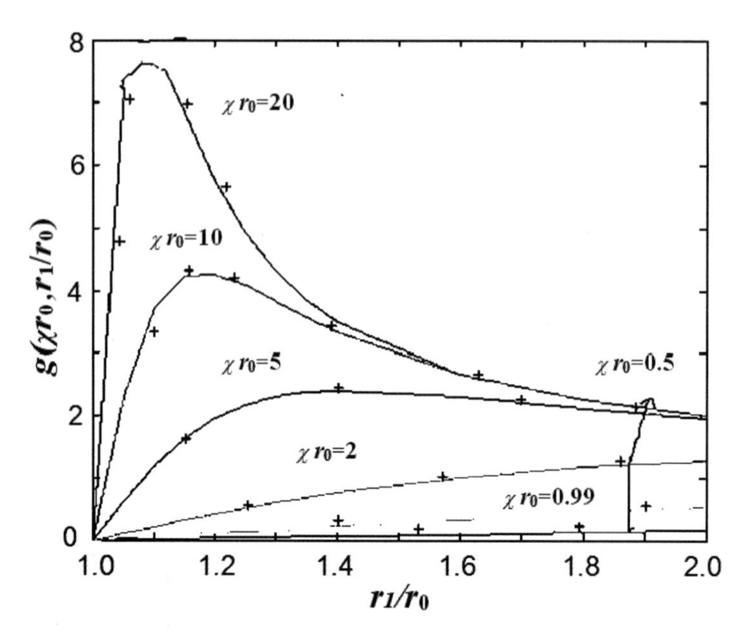

Figure 9.11. Dimensionless current $g(\chi r_0, r_1/r_0)$ as a function of r_1/r_0 for different enzyme reaction/substrate diffusion ratios χr_0. Key: (---) represents Eq. (9.74) and (+) represents Eq. (9.76) (Phanthong and Somasundrum 2003).

when $r_1/r_0 \geq 1.2$ for all values of χr_0. Finally, the current becomes equal value when $\chi r_0 \geq 5$ and $r_1/r_0 \geq 1.7$.

3.4. ZERO-ORDER CATALYTIC KINETICS

In this case, the substrate concentration c_S is greater than the Michaelis constant K_M. Now Eqs. (9.63) and (9.64) reduce to

$$\frac{D_S}{r^2}\frac{d}{dr}\left(r^2\frac{dc_S}{dr}\right) - k_{\text{cat}}c_E = 0 \tag{9.80}$$

$$\frac{D_H}{r^2}\frac{d}{dr}\left(r^2\frac{dc_H}{dr}\right) + k_{\text{cat}}c_E = 0 \tag{9.81}$$

The solution of the above equations using the boundary condition (9.65) becomes

$$C_S = \frac{r_1^2}{r_0^2} - \frac{r^2}{r_0^2} + 2\frac{r_0}{r_1} - 2\frac{r_0}{r} \tag{9.82}$$

$$C_H = \frac{r^2}{r_0^2} - \frac{r_1}{r_0} - \frac{r_1^2}{r_0^2} + \frac{r_1}{r} + \frac{r_1^2}{r_0 r} - 1 \tag{9.83}$$

where

$$C_S = \frac{6(c_S - c_S^*)}{(\chi r_0)^2 K_M} \tag{9.84}$$

and

$$C_H = \frac{c_H D_H}{(\chi r_0)^2 D_S K_M} \tag{9.85}$$

The current is

$$\frac{I r_0}{(\chi r_0)^2 nFAc_S^* D_S} = \frac{r_1}{r_0} + \frac{r_1^2}{r_0^2} - 2 \tag{9.86}$$

3.4.1. Discussion when $c_S > K_M$

Equations (9.82) and (9.83) represent the most general new analytical expressions for the substrate and hydrogen peroxide concentration profile for all values of r_1/r_0. In Figure 9.12 we present the series of normalized concentration profiles for a substrate as a function of the dimensionless parameters for various values

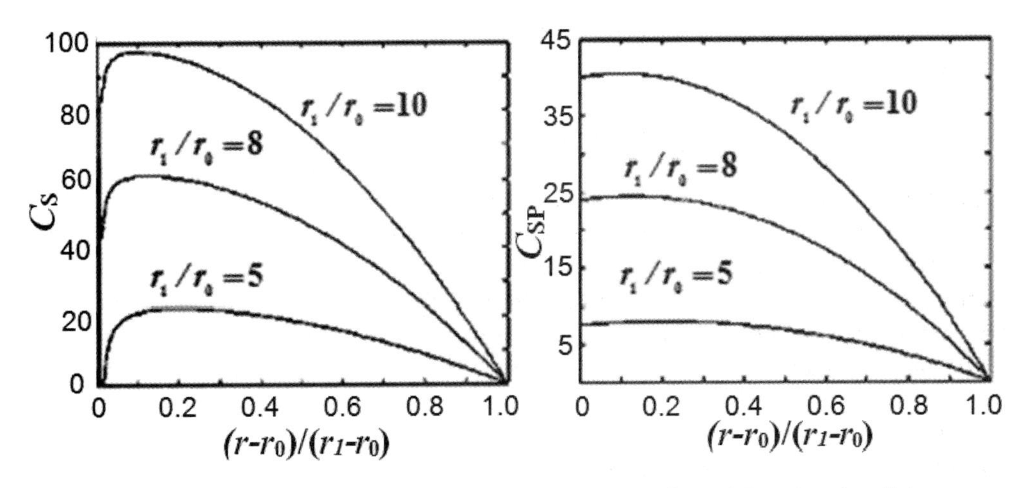

Figure 9.12. Dimensionless concentration profiles of substrates C_S and C_{SP} at a microdisk.

of film thickness r_1/r_0. From this figure it is inferred that the value of the concentration increases when the value of r_1/r_0 increases. Then the concentration C_S becomes zero when $r/r_1 = 1$ for all values of r_1/r_0.

3.5. FOR ALL VALUES OF K_M

In this case the homotopy perturbation method (He 1998, 2005, 2003b, 2006b; Ganji 2006) is used to give approximate analytical solutions of coupled nonlinear reaction diffusion equations (9.63) and (9.64). Using the homotopy perturbation method the approximate solutions of Eqs. (9.63) and (9.64) are

$$C_{SP} = \frac{r}{r_0} - 0.5\frac{r^2}{r_0^2} + 0.5\frac{r_1^2}{r_0^2} - \frac{r_1}{r_0} \tag{9.87}$$

$$C_{HP} = \frac{1}{2}\left(\frac{r_1 r}{r_0^2} - \frac{r^2}{r_0^2} - \frac{r_1}{r_0} + \frac{r}{r_0}\right) \tag{9.88}$$

where

$$C_{SP} = \frac{(c_S - c_S^*)}{c_S^*(\chi r_0)^2} \quad \text{and} \quad C_{HP} = \frac{c_H D_H}{(\chi r_0)^2 D_S c_S^*} \tag{9.89}$$

Equations (9.87) and (9.88) satisfy the boundary condition (9.65). Using Eq. (9.66), we obtain the current as

$$\frac{I r_0}{(\chi r_0)^2 nFAc_S^* D_S} = \frac{r_1 - r_0}{2r_0} \tag{9.90}$$

3.5.1. Discussion for All Values of c_S and K_M

Equations (9.87) and (9.88) represent the most general new approximate analytical expressions for the substrate and hydrogen peroxide concentration profile for all values of r_1/r_0. In Figure 9.13 we present the series of normalized concentration profiles for a substrate as a function of the dimensionless parameters for various values of film thickness r_1/r_0. From this figure it is concluded that the value of the concentration increases when the value of r_1/r_0 increases. Then the concentration C_{SP} becomes zero when $r/r_1 = 1$ for all values of r_1/r_0. Figure 9.13 shows the dimensionless concentration profiles of hydrogen peroxide C_{HP} for all values of r_1/r_0. From this figure it is deduced that the value of the concentration increases when the value of r_1/r_0 increases.

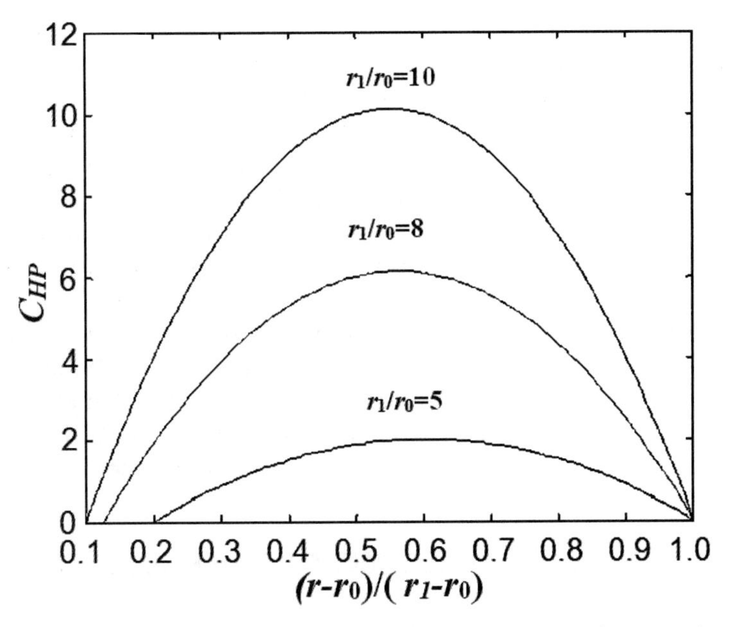

Figure 9.13. Dimensionless concentration profile of hydrogen peroxide C_{HP} at a microdisk.

3.6. CONCLUSIONS

A linear time-independent partial differential equation has been solved analytically. Analytical expressions for the steady-state substrate concentration and current have been derived by using the reduction of order when c_S is less than K_M. The concentration and current for enzyme reaction for all values of K_M have also been derived using the homotopy perturbation method. Good agreement with the available limiting case results is thus verified.

4. MICROCYLINDER BIOSENSORS

4.1. INTRODUCTION

Microelectrodes are used in biosensors (Revzin et al. 2002; Shi et. al 2002; Gue et al. 2002). This is due to factors such as fast response times, high signal:noise ratios, and the ability to operate in low-conductivity media, submicro volumes, and in vivo (Edmonds et al. 1985). The most commonly used microelectrodes in biosensors are microcylinders such as carbon fibers. This is because they are cheap, readily available, their form is suited to implantation (Gonon et al. 1992), and because much is known about their surface characteristics (Donnet and Basal 1984).

Among all the enzyme immobilization methods, the layer-by-layer (LbL) self-assembly process is a simple technique which may be applied to a wide range of enzymes (Decher and Hong 1991). This property is important both for constructing and for modeling studies of biosensors. The layer-by-layer process was first introduced by Decher and Hong (Decher and Hong 1991). This method has been applied to planar gold electrodes (Hodak et al. 1997; Forzani et al. 2000), carbon electrodes (Coche-Guerante et al. 2001), and polystyrene latex (Lvov and Caruso 2001; Fang et al. 2002; Caruso and Schuler 2000; Caruso et al. 2000; Sun and Hu et al. 2004).

Recently, Rijiravanich et al. (2006) obtained the steady-state concentration profile of *o*-quinone and dimensionless sensor response j for the limiting cases of low substrate concentrations. Even more recently, Venugopal et al. (2011) derived new and simple analytical solutions of the concentration and the current for all values of parameters using the homotopy perturbation method.

4.2. MATHEMATICAL FORMULATION OF THE PROBLEM

The model of a cylindrical electrode modified with both an enzyme and conducting sites/particles (circles) is shown in Figure 9.14. The cylindrical electrode is

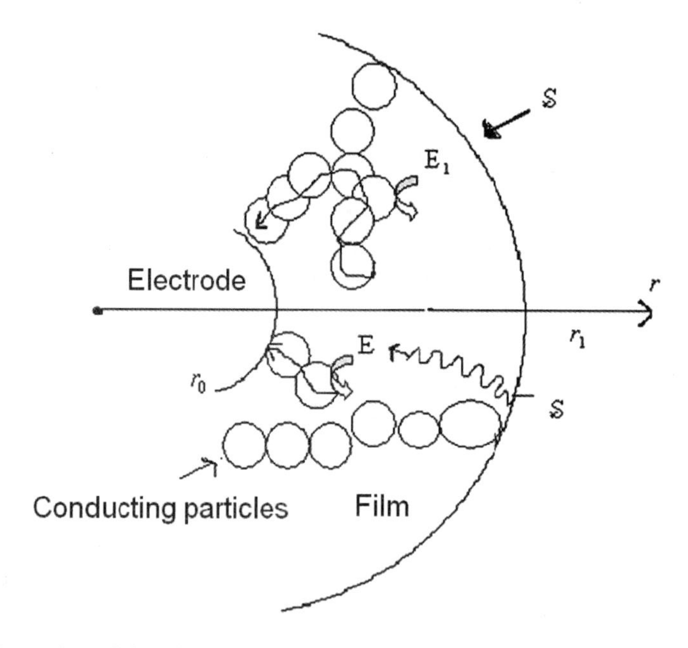

Figure 9.14. Illustration of the model of a cylindrical electrode modified with both an enzyme and conducting sites/particles (circles). (Reprinted with permission from Eswari and Rajendran 2010. Copyright 2010 Elsevier.)

uniformly coated by an enzyme immobilized in nonconducting material which is porous to the substrate. The electrode is used in a stirred solution containing an excess of supporting electrolyte. The enzyme and electrode reaction are (Rijiravanich et al. 2006)

$$O_2 + 2\ \text{catechol} \rightarrow 2\ o\text{-quinone} + 2\ H_2O \tag{9.91}$$

$$o\text{-quinone} + 2\ H^+ + 2\ e^- \rightarrow \text{catechol} \tag{9.92}$$

It is assumed that the enzyme concentration is uniform and that the enzyme reaction follows Michaelis-Menten kinetics, in which case the reaction in the film is (Carbanes et al. 1987)

$$S + E_1 \underset{k_2}{\overset{k_1}{\Leftrightarrow}} [E_1 S] \xrightarrow{k_{cat}} P + E_2 \tag{9.93}$$

where

$$k_{cat} = k_1 c_{O_2} \qquad \text{and} \qquad K_M = \frac{k_1 (k_2 + k_3) c_{O_2}}{k_2 k_3} \tag{9.94}$$

are the rate constant and the Michaelis-Menten constant. The mass balance for catechol, c_C, can be written in cylindrical coordinates as

$$\frac{D_C}{r}\frac{d}{dr}\left(r\frac{dc_C}{dr}\right) - \frac{k_{cat} c_E c_C}{c_C + K_M} = 0 \tag{9.95}$$

where c_C is the concentration profile of catechol, c_E is the concentration profile of enzyme, D_C and D_Q are its diffusion coefficients, K_M is the Michaelis constant, and c_Q is the concentration profile of quinone. Then the equation of continuum for quinone is generally expressed in the steady state by (Rijiravanich et al. 2006)

$$\frac{D_Q}{r}\frac{d}{dr}\left(r\frac{dc_Q}{dr}\right) + \frac{k_{cat} c_E c_C}{c_C + K_M} = 0 \tag{9.96}$$

At the electrode surface (r_0) and at the film surface (r_1) the boundary conditions are given by (Rijiravanich et al. 2006)

$$\begin{aligned} r = r_0: &\quad c_C = c_C^* \quad c_Q = 0 \\ r = r_1: &\quad c_C = c_C^* \quad c_Q = 0 \end{aligned} \tag{9.97}$$

where c_C^* is the bulk concentration of catechol scaled by the partition coefficient of the enzyme film. Adding Eqs. (9.95) and (9.96) and integrating with boundary condition (9.97) yields

$$\frac{c_C(r)}{c_C^*} + \frac{D_Q c_Q(r)}{D_C c_C^*} = 1 \tag{9.98}$$

The steady-state current can be given as (Rijiravanich et al. 2006)

$$\frac{I}{nF} = 2\pi L r_0 D_Q \left(\frac{dc_Q}{dr}\right)_{r=r_0} \tag{9.99}$$

4.3. ANALYTICAL SOLUTIONS OF THE CONCENTRATIONS AND CURRENT

Recently, Meena and Rajendran (2010), Anitha et al. (2010b), and Manimozhi et al. (2010) implemented the homotopy perturbation method to give approximate and analytical solutions of nonlinear reaction-diffusion equations containing a nonlinear term related to the Michaelis-Menten kinetics of the enzymatic reaction. Eswari et al. (2010a) solved the coupled nonlinear diffusion equations analytically for microdisk and microcylinder enzyme electrodes when a product from an immobilized enzyme reacts with the electrode. Using the homotopy perturbation method, the dimensionless concentration of the catechol and *o*-quinone are

$$C(R) = 1 + \left[\frac{\gamma_E R^2 - \gamma_E\left(1 + r_1/r_0\right)R + \gamma_E\left(r_1/r_0\right)}{2(1+\alpha)}\right] \tag{9.100}$$

$$Q(R) = \left[\frac{-\gamma_S R^2 + \gamma_S\left(1 + r_1/r_0\right)R - \gamma_S\left(r_1/r_0\right)}{2(1+\alpha)}\right] \tag{9.101}$$

where

$$C = \frac{c_C}{c_C^*} \quad Q = \frac{c_Q}{c_C^*} \quad R = \frac{r}{r_0} \quad \alpha = \frac{c_C^*}{K_M} \quad \gamma_E = \frac{k_{cat} c_E r_0^2}{D_C K_M} \quad \gamma_S = \frac{k_{cat} c_E r_0^2}{D_Q K_M}$$

$$\frac{D_Q}{D_C} = \frac{\gamma_E}{\gamma_S} \tag{9.102}$$

Equations (9.100) and (9.101) satisfy the boundary conditions (9.97). These equations represent the new and simple analytical expression of the concentration of catechol and o-quinone for all possible values of the parameters γ_E, γ_S, α, and r_1/r_0. Equations (9.100) and (9.101) also satisfy the relation $C(R) + (\gamma_E/\gamma_S)Q(R) = 1$. From Eqs. (9.100) and (9.101), we can obtain the dimensionless current, which is as follows:

$$\psi = I/nFL\,D_Q c_C^* = 2\pi\left[\frac{\gamma_S\left(1+r_1/r_0\right)-2\gamma_S}{2(1+\alpha)}\right] \tag{9.103}$$

Equation (9.103) represents the new and closed form of an analytical expression for the current for all possible values of parameters.

4.3.1. Limiting Cases for Unsaturated (First-Order) Catalytic Kinetics

In this case, the catechol concentration c_C is less than the Michaelis constant K_M. Now Eqs. (9.95) and (9.96) reduce to the following forms:

$$\frac{D_C}{r}\frac{d}{dr}\left(r\frac{dc_C}{dr}\right)-\frac{k_{\text{cat}}c_E c_C}{K_M}=0 \tag{9.104}$$

$$\frac{D_Q}{r}\frac{d}{dr}\left(r\frac{dc_Q}{dr}\right)+\frac{k_{\text{cat}}c_E c_C}{K_M}=0 \tag{9.105}$$

By solving Eq. (9.104) using the boundary condition [Eq. (9.97)], the concentration of catechol c_C can be obtained in the form of modified Bessel functions of zeroth order $I_0(\chi r)$ and $K_0(\chi r)$.

$$\frac{c_C(r)}{c_C^*}=\left\{\frac{I_0(\chi r)\,[K_0(\chi r_0)-K_0(\chi r_1)]+K_0(\chi r)\,[I_0(\chi r_1)-I_0(\chi r_0)]}{K_0(\chi r_0)I_0(\chi r_1)-K_0(\chi r_1)I_0(\chi r_0)}\right\} \tag{9.106}$$

where $\chi^2 = k_{\text{cat}}c_E/D_C K_M$. Inserting Eq. (9.106) into Eq. (9.98), we can obtain the concentration $c_Q(r)$:

$$\frac{D_Q c_Q(r)}{D_C c_C^*}=1-\left\{\frac{I_0(\chi r)\,[K_0(\chi r_0)-K_0(\chi r_1)]+K_0(\chi r)\,[I_0(\chi r_1)-I_0(\chi r_0)]}{K_0(\chi r_0)I_0(\chi r_1)-K_0(\chi r_1)I_0(\chi r_0)}\right\} \tag{9.107}$$

The sensor response j in terms of modified Bessel function of zeroth order can be obtained as follows:

$$j=\frac{I}{nFLD_cC_c^*}=\frac{2\pi\chi r_0\left(K_1(\chi r_0)\,[I_0(\chi r_1)-I_0(\chi r_0)]-I_1(\chi r_0)\,[K_0(\chi r_0)-K_0(\chi r_1)]\right)}{K_0(\chi r_0)I_0(\chi r_1)-K_0(\chi r_1)I_0(\chi r_0)} \tag{9.108}$$

Table 9.2. Values of *p* and *q* which fit Eq. (9.111) to Eq. (9.110) with <5% error and *a* valid for $\alpha_1 \leq 2.0$ and *b* valid for $\alpha_1 > 2.0$

x	*p*	*q*
9.0–7.0	1.00	1.01
6.0–4.0	1.03	1.05
3.0	1.04	1.10
2.0	1.02	$1.14^a/1.25^b$

Table 9.2 shows the values of *p* and *q* which fit Eq. (9.111) to Eq. (9.110) with <5% error and *a* valid for $\alpha_1 \leq 2.0$ and *b* valid for $\alpha_1 > 2.0$.

4.4. COMPARISON WITH LIMITING CASE OF RIJIRAVANICH'S WORK

Recently, Rijiravanich et al. (2006) derived an analytical expression of the steady-state concentration c_Q [Eq. (9.109)] and sensor response *j* [Eqs. (9.110) and (9.109)] in integral form for the limiting case $c_C < K_M$.

$$\frac{D_Q C_Q(r)}{D_c C_c^*} = g\chi\left\{-f\int_{r_0}^{r} I_1(\chi r)\,dr + \int_{r_0}^{r} K_1(\chi r)\,dr + \frac{\ln(r/r_0)}{\ln(r_1/r_0)}\left[f\int_{r_0}^{r_1} I_1(\chi r)\,dr - \int_{r_0}^{r_1} K_1(\chi r)\,dr\right]\right\} \tag{9.109}$$

Table 9.3. Comparison of dimensionless sensor response *j* for various values of χr_0 using Eqs. (9.109) and (9.112) when the thickness of the film $\alpha_1 = r_1/r_0 = 5$

$x(=\chi r_0)$	$\alpha_1 = r_1/r_0$	*p*	*q*	Eq. (9.112) (Rijiravanich et al. 2006)	Eq. (109) (this work)	Error (%)
9	5	1	1.01	57.78	57.78	0.00
8	5	1	1.01	51.30	51.30	0.00
7	5	1	1.01	44.82	44.78	0.09
5	5	1.03	1.05	34.03	34.01	0.06
4	5	1.03	1.05	26.92	25.95	3.77
3	5	1.04	1.10	21.03	20.99	0.19
2	5	1.02	1.25	14.93	14.93	0.01
Average percent deviation						0.59

Table 9.4. Comparison of dimensionless sensor response j for various values of χr_0 using Eqs. (9.108) and (9.111) when the thickness of the film $\alpha_1 = r_1/r_0 = 1.5$

$x(=\chi r_0)$	$\alpha_1 = r_1/r_0$	p	q	Eq. (9.111) (Rijiravanich et al. 2006)	Eq. (9.108) (this work)	Error (%)
9	1.5	1	1.01	56.51	56.51	0.00
8	1.5	1	1.01	49.45	49.45	0.00
7	1.5	1	1.01	42.20	42.19	0.02
5	1.5	1.03	1.05	28.62	27.60	3.67
4	1.5	1.03	1.05	20.27	20.43	0.80
3	1.5	1.04	1.10	13.09	13.15	0.45
2	1.5	1.02	1.14	6.32	6.32	0.00
Average percent deviation						0.71

$$j = \frac{I}{nFLD_c C_c^*} = 2\pi\chi g X \left\{ r_0 \left[fI_1(\chi r_0) + K_1(\chi r_0) \right] + \frac{1}{\ln(r_1/r_0)} \left[f \int_{r_0}^{r_1} I_1(\chi r)\, dr - \int_{r_0}^{r_1} K_1(\chi r)\, dr \right] \right\} \tag{9.110}$$

where $g = 1/[fI_0(\chi r_0) + K_0(\chi r_0)]$, $f = [K_0(\chi r_0) - K_0(\chi r_1)]/[I_0(\chi r_1) - I_0(\chi r_0)]$. Rijiravanich et al. (2006) obtained the empirical expression of the current as

$$j = 2\pi x^q \tanh\left[\left(\frac{x}{2} \right) (\alpha_1 - 1) \right]^p \tag{9.111}$$

where p and q are empirical constants and $\alpha_1 = r_1/r_0$. The values of p and q are given for various values of $x(=\chi r_0)$ in Tables 9.2–9.4. This empirical expression is compared to our closed analytical expression [Eq. (9.109)] in Tables 9.3 and 9.4. The average relative difference between our Eq. (9.109) and the empirical expression Eq. (9.111) is 0.71% when $\alpha_1 = 1.5$ and 0.59% when $\alpha_1 = 5$.

Table 9.3 represents the comparison of dimensionless sensor response j for various values of χr_0 using Eqs. (9.109) and (9.112) when the thickness of the film is $\alpha_1 = r_1/r_0 = 5$.

Table 9.4 shows the comparison of dimensionless sensor response j for various values of χr_0 using Eqs. (9.108) and (9.111) when the thickness of the film $\alpha_1 = r_1/r_0 = 1.5$).

4.5. DISCUSSION

The dimensionless concentration profile of catechol, $C(R)$, using Eq. (9.100) for all various values of the parameters γ_S, γ_E, r_1/r_0, and α is plotted in Figure 9.15. It is

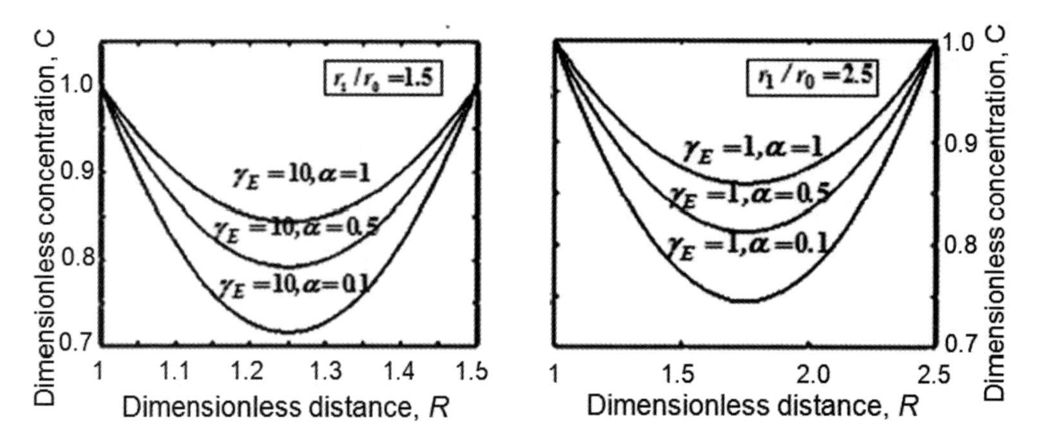

Figure 9.15. Typical normalized steady-state concentration profile of catechol $C(R)$ plotted from Eq. (9.100) for different values of parameters γ_E and α when $r_1/r_0 = 1.5$ and $r_1/r_0 = 2.5$.

concluded that there is a simultaneous increase in the values of the concentration of catechol as well as in the saturated parameter α for small values of γ_E. Also, the value of catechol concentration C is approximately equal to 1 when $R = 1$ and $R = r_1/r_0$ for all values of α and γ_E. Figure 9.16 shows the concentration profile of *o*-quinone, $Q(R)$, in R space for various values of α and γ_S calculated using Eq. (9.101). The plot was constructed for $r_1/r_0 = 1.5$ and 5. From these figures, it is confirmed that the value of the concentration of *o*-quinone increases when $\gamma_S \geq 0.1$ for small values of α. From Figures 9.15 and 9.16, we can observe that the dimensionless concentration of catechol should vary between 0 and 1. Because catechol is converted to *o*-quinone, the *o*-quinone concentration should be the inverse of that

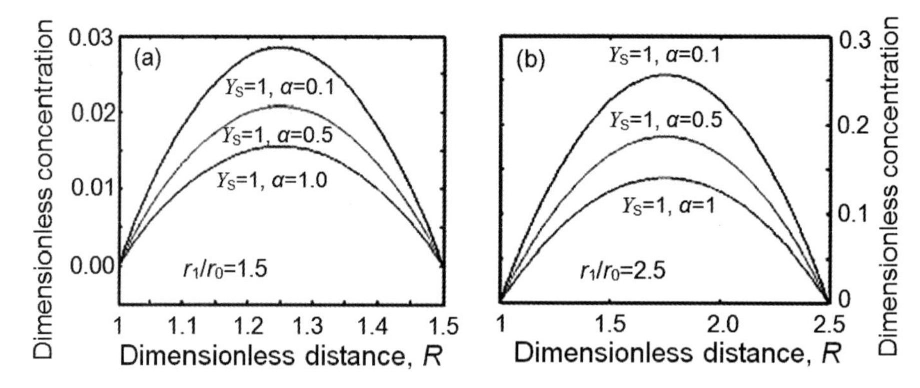

Figure 9.16. Typical normalized steady-state concentration profile of $Q(R)$ plotted from Eq. (9.101) for different values of parameters γ_S and α when $r_1/r_0 = 1.5$ and $r_1/r_0 = 2.5$.

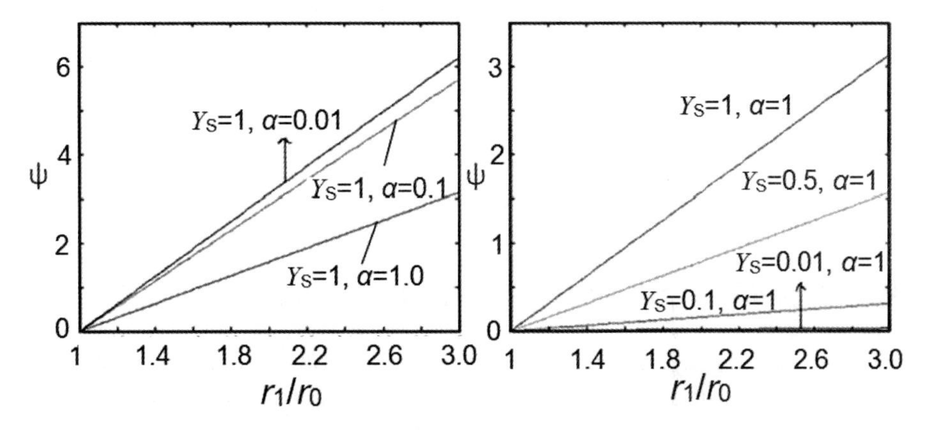

Figure 9.17. Plot of dimensionless current ψ versus r_1/r_0. Current is calculated by Eq. (9.103).

of catechol. The substrate catechol C is minimum and product o-quinone Q is maximum when $R = (0.5 + r_1/2r_0)$ for all values of γ_S and α. The minimum value of concentration of catechol is

$$C_{\min} = \frac{8 + 8\alpha - \gamma_E + 2\gamma_E\alpha_1 - \gamma_E\alpha_1^2}{8(1+\alpha)} \tag{9.112}$$

and the maximum value of the concentration profile of quinone is

$$Q_{\max} = \frac{\gamma_S(1 - 2\alpha_1 + \alpha_1^2)}{8(1+\alpha)}$$

where $r_1/r_0 = \alpha_1$. The dimensionless current ψ versus r_1/r_0 using Eq. (9.103) is plotted in Figure 9.17. The value of current ψ increases when the thickness of the film r_1/r_0 and the dimensionless reaction-diffusion parameter γ_S either increases or decreases.

4.6. CONCLUSIONS

The primary result of this work is simple approximate calculation of concentration of catechol, o-quinone, and current for all values of parameters γ_E, γ_S, α, r_1/r_0, and χr_0. Formerly, Rijiravanich et al. (2006) have considered the first-order kinetics (linear range) only. However, in this paper, the most important nonlinear contributions of biosensors are reported.

The length of the linear range is also an important analytical parameter. In developing a sensor, experimental scientists would like this range to cover all concentrations expected in actual samples, as this makes calibration of the sensor in the field much easier. In Tables 9.3 and 9.4, our analytical results are compared with the limiting case of first-order catalytic kinetics (Rijiravanich et al. 2006) results, which yield good agreement with the previous limiting-case results.

4.7. PPO-MODIFIED MICROCYLINDER BIOSENSORS

Venugopal et al. (2011a) have developed mathematical model for steady state current in phenol–polyphenol oxidase (PPO)–modified microcylinder biosensors. It is assumed that the enzyme concentration is uniform and that the enzyme reaction follows Michaelis-Menten kinetics:

$$S + E_1 \underset{k_2}{\overset{k_1}{\Leftrightarrow}} [E_1S] \xrightarrow{k_{cat}} P + E_2$$

The mass balance for catechol can be written in cylindrical coordinates as follows:

$$\frac{D_C}{r}\frac{d}{dr}\left(r\frac{dC_c}{dr}\right) - \frac{k_{cat}C_E C_c}{C_c + k_M} = 0 \quad (9.113)$$

Then the equation of the continuum for quinone is generally expressed in the steady state as

$$\frac{D_Q}{r}\frac{d}{dr}\left(r\frac{dC_Q}{dr}\right) - \frac{k_{cat}C_E C_c}{C_c + k_M} = 0 \quad (9.114)$$

where C is the concentration profile of catechol, C_E is the concentration profile of enzyme, D_C and D_Q are diffusion coefficients, k_M is the Michaelis constant, and C_Q is the concentration profile of quinone. Boundary conditions are given by $r = r_0; C_c = C_c^*; C_Q = 0$; $r = r_1; C_c = C_c^*; C_Q = 0$, where (r_0) is the electrode surface and (r_1) is the film surface. Now the dimensionless parameters are defined as

$$C = \frac{C_C}{C_C^*} \quad Q = \frac{C_Q}{C_C^*} \quad R = \frac{r}{r_0} \quad \alpha = C_C^* \quad \gamma_E = \frac{k_{cat}C_E r_0^2}{D_C K_M} \quad \gamma_S = \frac{k_{cat}C_E r_0^2}{D_Q K_M} \quad \frac{D_Q}{D_C} = \frac{\gamma_E}{\gamma_S} \quad (9.115)$$

where C_C^* is the bulk concentration of catechol scaled by the partition coefficient of the enzyme film. The dimensionless equation is of the form

$$\frac{d^2C}{dR^2}+\frac{1}{R}\frac{dC}{dR}-\frac{\gamma_E C}{1+\alpha C}=0$$
$$\frac{d^2Q}{dR^2}+\frac{1}{R}\frac{dQ}{dR}-\frac{\gamma_S C}{1+\alpha C}=0 \qquad (9.116)$$

The boundary conditions are represented as follows: $C = 1$, $Q = 0$ when $R = 1$; $C = 1$, $Q = 0$ when $R = r_1/r_0$. The analytical expressions of concentration of catechol and enzyme are as follows:

$$C(R)=1+\left[\frac{\gamma_E R^2-\gamma_E(1+r_1/r_0)R+\gamma_S(r_1/r_0)}{2(1+\alpha)}\right]$$
$$Q(R)=\left[\frac{-\gamma_S R^2+\gamma_S(1+r_1/r_0)R-\gamma_S(r_1/r_0)}{2(1+\alpha)}\right] \qquad (9.117)$$

5. SPHERICAL BIOSENSORS

5.1. SIMPLE MICHAELIS-MENTEN AND PRODUCT COMPETITIVE INHIBITION KINETICS

Angel Joy et al. (2011) developed a two-parameter mathematical model for immobilized enzymes and the homotopy analysis method. The coupled differential equations for substrate and product in spherical coordinates are

$$\frac{D_S}{r^2}\frac{d}{dr}\left(r^2\frac{dC_S}{dr}\right)=V_S$$
$$\frac{D_P}{r^2}\frac{d}{dr}\left(r^2\frac{dC_P}{dr}\right)=-V_S \qquad (9.118)$$

The boundary conditions are, when $r = 0$, $dC_S/dr = 0$; $dC_P/dr = 0$; at $r = R$, $C_S = C_{SR}$; $C_P = C_{PR}$ and reaction rate $V_S = V_m\left[C_S-\left(C_P/K_{eq}\right)\right]/\left[K_m+C_S+\left(K_m/K_P\right)C_P\right]$. Here C_S and C_P denote the dimensional substrate and product concentration, r is the radial coordinate, R denotes the radius of the particle, D_S and D_P are the diffusion coefficients of the substrate and product, respectively, C_{SR} and C_{PR} denote the local substrate and product concentration, K_{eq} is the reaction equilibrium constant, K_m is the Michaelis-Menten constant, and V_m defines the maximum reaction rate. The following relationship can be established:

$$C_P=C_{PR}+\frac{D_S}{D_P}\left(C_{SR}-C_S\right)$$

Substituting the value of C_P in V_S, we can obtain

$$V_S = \frac{V_m\left[1+\left(1/K_{eq}\right)\left(D_S/D_P\right)\right]\left(C_S - C_{SE}\right)}{K_M + \left(K_M/K_P\right)C_{PE} + C_{SE} + \left[1-\left(K_M/K_P\right)\left(D_S/D_P\right)\right]\left(C_S - C_{SE}\right)} \quad (9.119)$$

where

$$K_{eq} = \frac{C_{PE}}{C_{SE}} \qquad C_{SE} = \frac{C_{PR} + \left(D_S/D_P\right)C_{PR}}{K_{eq} + \left(D_S/D_P\right)} \qquad C_{PE} = K_{eq}C_{SE} = \frac{C_{PR} + \left(D_S/D_P\right)C_{PR}}{1+\left(1/K_{eq}\right)\left(D_S/D_P\right)} \quad (9.120)$$

We introduce the following dimensionless parameters:

$$U = \frac{C_S - C_{SE}}{C_{SR} - C_{SE}} \qquad \rho = \frac{r}{R} \qquad \phi = \frac{R^2 V_m}{\left(C_{SR} - C_{SE}\right)D_S}\frac{\left[1+\left(1/K_{eq}\right)\left(D_S/D_P\right)\right]}{\left[1-\left(K_M/K_P\right)\left(D_S/D_P\right)\right]}$$

$$\alpha = \frac{K_M + \left(K_M/K_P\right)C_{PE} + C_{SE}}{\left(C_{SR} - C_{SE}\right)\left[1-\left(K_M/K_P\right)\left(D_S/D_P\right)\right]} \quad (9.121)$$

where U represents the dimensionless substrate concentration, ρ denotes the dimensionless radius of the particle, ϕ and α denote the dimensionless modulus. Then the differential equation in dimensionless form is

$$\frac{1}{\rho^2}\frac{d}{d\rho}\left(\rho^2\frac{dU}{d\rho}\right) = \phi\frac{U}{\alpha + U} \quad (9.122)$$

The boundary conditions are $\rho = 0$, $dU/d\rho = 0$; and $\rho = 1$, $U = 1$. The effectiveness factor is

$$\eta = 3(\alpha+1)\int_0^1 \frac{U}{U+\alpha}\rho^2\, d\rho \quad (9.123)$$

The new approximate substrate concentration using the homotopy analysis method is

$$U(\rho) = 1 - \frac{h\phi\left(\rho^2 - 1\right)}{6\alpha} + \frac{h\phi\left\{(h\phi/20)\rho^4 - \left[(1+h)\alpha + h + (\phi h/6)\right]\rho^2 + \left[(1+h)\alpha + h + (7h\phi/60)\right]\right\}}{6\alpha^2} \quad (9.124)$$

The effectiveness factor is

$$\eta = \frac{(\alpha+1)}{\phi h A}\left[\phi h A + 18\alpha^2 (A - \phi h B) - 108\alpha^3 B(\alpha+1)\right] \tag{9.125}$$

where

$$A = \sqrt{\phi^2 h^2 + 6\phi h\alpha(\alpha+1)} \qquad \text{and} \qquad B = \operatorname{arctan} h\left(\frac{\phi h}{A}\right) \tag{9.126}$$

5.2. IMMOBILIZED ENZYME FOR SPHERICAL BIOSENSORS

Margret PonRani et al (2011) developed an analytical expression for the substrate concentration in different parts of particles with immobilized enzyme and substrate inhibition kinetics. The nonlinear differential equation described by the concentration within the enzymatic layer at steady-state conditions is

$$\nabla^2 C_S - \frac{V_m C_S}{K_m + C_S + \left(C_S^2/k_i\right)} = 0 \tag{9.127}$$

where C_S is the substrate concentration, K_m is the Michaelis constant, K_i is the substrate inhibition constant, and V_m is the maximum reaction rate. Here ∇^2 stands for the Laplace operator. The boundary conditions $\partial C_S/\partial r = 0$ when $r = 0$ and $C_S = C_{SO}$ when $r = R_P$, where C_{SO} is the substrate concentration on the particle surface. The dimensionless parameters are defined by

$$C = \frac{C_S}{K_m} \qquad x = \frac{r}{R_P} \qquad \Phi = R_P\sqrt{\frac{V_m}{K_m D_e}} \qquad \beta = \frac{k_i}{k_m} \qquad m = \frac{C_{SO}}{K_m} \tag{9.128}$$

where C is the dimensionless substrate concentration, x is the dimensionless particle radial coordinate, Φ is the Thiele modulus, β is the dimensionless kinetic parameter, and m is the dimensionless substrate concentration on the particle surface. The reduced dimensionless form for a cylindrical particle is

$$\frac{\partial^2 C}{\partial x^2} + \frac{1}{x}\frac{\partial C}{\partial x} - \Phi^2 \frac{C}{1 + C + \left(C^2/\beta\right)} = 0 \tag{9.129}$$

The boundary conditions are $\partial C / \partial x = 0$ when $x = 0$ and $C = m$ when $x = 1$. The analytical solution of the above equation is

$$C = m + \frac{\Phi^2 \beta m}{2(\beta + \beta m + m^2)}(x^2 - 1) \tag{9.130}$$

5.3. CONCLUSION

Several experimental problems relevant to ongoing research in this area have been presented, which were mostly concerned with optimizing design parameters for biosensing devices. We constructed mathematical models for these problems and, using analytical methods or numerical simulations, attempted to describe the behavior of solutions, with a view to providing recommendations for improving experimental practice. As well as studying these practical problems directly, a large part of the chapter was dedicated to reviewing existing mathematical models relevant to biosensor design, usually involving kinetics or transport of chemical species. This work can therefore be used to provide simplified modeling strategies for chemistry researchers. We have also examined the connection between modeling results and experimental calibration curves as well as the possibility of tracking no-specific biosensor responses.

APPENDIX: VARIOUS ANALYTICAL SCHEMES FOR SOLVING NONLINEAR REACTION DIFFUSION EQUATIONS

A. BASIC CONCEPT OF THE VARIATIONAL ITERATION METHOD

The variational iteration method (He et al. 2010; He and Wu 2007; Lu Junfeng 2007; Abdou et al. 2005) has been extensively worked out over a number of years by numerous authors. The variational iteration method has been favorably applied to various kinds of nonlinear problems (Lu Junfeng 2007; Abdou et al. 2005). The main property of the method is its flexibility and ability to solve nonlinear equations (He and Wu 2007). Recently, Rahamathunissa and Rajendran (2008), Senthamarai and Rajendran (2010), and Meena et al. (2010) have implemented the variational iteration method to give approximate and analytical solutions of nonlinear reaction diffusion equations containing a nonlinear term related to the Michaelis-Menten kinetics of the enzymatic reaction. Besides its mathematical importance and its links to other branches of mathematics, it is widely used in all ramifications of modern sciences (He 2007). In this method the solution procedure is very simple by means of variational theory, and only a few iterations lead to highly accurate solutions which are valid for the whole solution domain. To illustrate the basic concept of the technique, we consider the following general differential equation:

$$L[u(x)] + N[u(x)] = g(x) \tag{9.A1}$$

where L is a linear operator, N is a nonlinear operator, and $g(x)$ is a given continuous function. Using the variational iteration method, we construct a correction functional as follows:

$$u_{n+1}(x) = u_n(x) + \int_0^x \lambda \{L[u_n(s)] + N[\tilde{u}_n(s)] - g(s)\} ds \tag{9.A2}$$

where λ is a Lagrangian multiplier which can be identified optimally via the variational iteration method. u_n is the nth approximate solution, $\tilde{u}_n$ denotes a restricted variation, i.e., $\delta\tilde{u}_n = 0$. Equation (9.A2) is called a correction functional. The solution of the linear problem can be solved in a single iteration due to the exact identification of the Lagrange multiplier. In this method, it is required to first determine the Langrange multiplier, λ, optimally. The successive approximation, $u_{n+1}(\chi)$, $n \geq 0$, of the solution of the correction functional will be readily obtained upon using the determined Lagrange multiplier and the first approximation, $u_0(\chi)$. Consequently, the solution is given by $u = \lim_{n\to\infty} u_n$.

B. BASIC CONCEPT OF THE HOMOTOPY PERTURBATION METHOD

Recently, many authors have applied the homotopy perturbation method (HPM) to various problems and demonstrated the efficiency of the HPM for handling nonlinear structures and solving various physics and engineering problems (Cai et al. 2006; Ariel 2010; Ghori et al. 2007; Ozis and Yildirim 2007; Li and Liu 2006; Mousa and Ragab 2008). This method is a combination of homotopy in topology and classic perturbation techniques. Ji-Huan He used the HPM to solve the Lighthill equation (He 1999), the Duffing equation (He 2003b), and the Blasius equation (He 2003a). The idea has been used to solve nonlinear boundary-value problems (He 2006a), integral equations (Golbabai and Keramati 2008; Ghasemi 2007; Biazar and Ghazvini 2009), Klein-Gordon and Sine-Gordon equations (Odibat 2007), Emden-Flower type equations (Chowdhury and Hashim 2007), and many other problems. This wide variety of applications shows the power of the HPM to solve functional equations. The HPM is unique in its applicability, accuracy, and efficiency. The HPM (He 2006b) uses the imbedding parameter p as a small parameter, and only a few iterations are needed to search for an asymptotic solution. The HPM has overcome the limitations of traditional perturbation methods. It can take full advantage of the traditional perturbation techniques, so a considerable amount of research has been conducted to apply the homotopy technique to solve various strong nonlinear equations (Cai et al. 2006; Ariel 2010). To explain this method, let us consider the following function:

$$D_o(u) - f(r) = 0 \qquad r \in \Omega \tag{9.B1}$$

with the boundary conditions

$$B_o\left(u, \frac{\partial u}{\partial n}\right) = 0 \qquad r \in \Gamma \tag{9.B2}$$

where D_o is a general differential operator, B_o is a boundary operator, $f(r)$ is a known analytical function, and Γ is the boundary of the domain Ω. Generally speaking, the operator D_o can be divided into a linear part L and a nonlinear part N. Equation (9.B1) can therefore be written as

$$L(u) + N(u) - f(r) = 0 \tag{9.B3}$$

By the homotopy technique, we construct a homotopy $v(r,p): \Omega \times [0,1] \rightarrow \Re$ that satisfies

$$H(v,p) = (1-p)[L(v) - L(u_0)] + p[D_o(v) - f(r)] = 0 \tag{9.B4}$$

$$H(v,p) = L(v) - L(u_0) + pL(u_0) + p[N(v) - f(r)] = 0 \tag{9.B5}$$

where $p \in [0,1]$ is an embedding parameter and u_0 is an initial approximation of Eq. (9.B1) that satisfies the boundary conditions. From Eqs. (9.B4) and (9.B5), we have

$$H(v,0) = L(v) - L(u_0) = 0 \tag{9.B6}$$

$$H(v,1) = D_o(v) - f(r) = 0 \tag{9.B7}$$

When $p = 0$, Eq. (9.B4) and Eq. (9.B5) become linear equations. When $p = 1$, they become nonlinear equations. The process of changing p from zero to unity is that of $L(v) - L(u_0) = 0$ to $D_o(v) - f(r) = 0$. We first use the embedding parameter p as a "small parameter" and assume that the solutions of Eqs. (9.B4) and (9.B5) can be written as power series in p:

$$v = v_0 + pv_1 + p^2 v_2 + \ldots \tag{9.B8}$$

Setting $p = 1$ results in the approximate solution of Eq. (9.B1):

$$u = \lim_{p \rightarrow 1} v = \mathrm{v}_0 + v_1 + v_2 + \ldots \tag{9.B9}$$

This is the basic idea of the HPM.

C. BASIC CONCEPT OF THE HOMOTOPY ANALYSIS METHOD

Liao (1992) proposed a powerful analytical method for nonlinear problems, namely, the homotopy analysis method (HAM). Different from all reported perturbation and nonperturbative techniques, the homotopy analysis method (Liao 2003, 2004; Domairry and Bararnia 2008; Awawdeh et al. 2009; Domairry and Fazeli 2009; Jafari et al. 2009; Sohouli et al. 2010) itself provides us with a convenient way to control and adjust the convergence region and rate of approximation series, when necessary. Briefly speaking, the homotopy analysis method has the following advantages: It is valid even if a given nonlinear problem does not contain any small (large) parameters at all; and it can be employed to efficiently approximate a nonlinear problem by choosing different sets of base functions. Consider the following differential equation:

$$N[u(t)] = 0 \tag{9.C1}$$

where N is a nonlinear operator, t denotes an independent variable, and $u(t)$ is an unknown function. For simplicity, we ignore all boundary or initial conditions, which can be treated in a similar way. By generalizing the conventional homotopy method, Liao constructed the so-called zero-order deformation equation as

$$(1-p)L[\phi(t;p) - u_0(t)] = phH(t)N[\phi(t;p)] \tag{9.C2}$$

where $p \in [0,1]$ is the embedding parameter, $h \neq 0$ is a nonzero auxiliary parameter, $H(t) \neq 0$ is an auxiliary function, L is an auxiliary linear operator, $u_0(t)$ is an initial guess of $u(t)$, and $\phi(t; p)$ is an unknown function. It is important that one has great freedom to choose auxiliary unknowns in the HAM. Obviously, when $p = 0$ and $p = 1$, it holds that:

$$\phi(t;0) = u_0(t) \qquad \text{and} \qquad \phi(t;1) = u(t) \tag{9.C3}$$

Thus, as p increases from 0 to 1, the solution $\phi(t; p)$ varies from the initial guess $u_0(t)$ to the solution $u(t)$. Expanding $\phi(t; p)$ in Taylor series with respect to p, we have

$$\phi(t;p) = u_0(t) + \sum_{m=1}^{+\infty} u_m(t)p^m \tag{9.C4}$$

where

$$u_m(t) = \frac{1}{m!}\frac{\partial^m \phi(t;p)}{\partial p^m}\bigg|_{p=0} \tag{9.C5}$$

If the auxiliary linear operator, the initial guess, the auxiliary parameter h, and the auxiliary function are properly chosen, the series (C4) converges at $p = 1$; then we have

$$u(t) = u_0(t) + \sum_{m=1}^{+\infty} u_m(t) . \tag{9.C6}$$

Define the vector

$$\vec{u}_n = \{u_0, u_1, \ldots, u_n\} \tag{9.C7}$$

Differentiating Eq. (9.C2) for m times with respect to the embedding parameter p, and then setting $p = 0$ and finally dividing them by $m!$, we will have the so-called mth-order deformation equation:

$$L[u_m - \chi_m u_{m-1}] = hH(t)\Re_m(\vec{u}_{m-1}) \tag{9.C8}$$

where

$$\Re_m(\vec{u}_{m-1}) = \frac{1}{(m-1)!} \left. \frac{\partial^{m-1} N[\phi(t;p)]}{\partial p^{m-1}} \right|_{p=0} \tag{9.C9}$$

and

$$\chi_m = \begin{cases} 0 & m \le 1 \\ 1 & m > 1 \end{cases} \tag{9.C10}$$

Applying L^{-1} on both side of Eq. (9.C8), we get

$$u_m(t) = \chi_m u_{m-1}(t) + hL^{-1}[H(t)\Re_m(\vec{u}_{m-1})] \tag{9.C11}$$

In this way, it is easy to obtain u_m. For $m \geq 1$, at Mth order, we have

$$u(t) = \sum_{m=0}^{M} u_m(t)$$

When $M \to +\infty$, we get an accurate approximation of the original equation (9.C1). For the convergence of the above method we refer the reader to Liao (2003). If Eq. (9.C1) admits a unique solution, then this method will produce the unique solution. If Eq. (9.C1) does not possess a unique solution, the HAM will give a solution among many other (possible) solutions.

D. BASIC CONCEPT OF THE ADOMIAN DECOMPOSITION METHOD

In recent years, much attention has been devoted to application of the Adomian decomposition method (ADM) to the solution of various scientific models (Adomian 1984; Patel and Serrano 2011; Mohamed 2010; Jaradat 2008; Siddiqui et al. 2010). The ADM yields, without linearization, perturbation, transformation, or discretization, an analytical solution in terms of a rapidly convergent infinite power series with easily computable terms. The Adomian decomposition method (Majid Wazwaz and Gorguis 2004; Biazar and Islam 2004; Sweilam and Khader 2010; Adomian 1994, 1995) depends on decomposing the nonlinear differential equation

$$F[\chi, y(\chi)] = 0 \tag{9.D1}$$

into the two components

$$L[y(\chi)] + N[y(\chi)] = 0 \tag{9.D2}$$

where L and N are the linear and the nonlinear parts of F, respectively. The operator L is assumed to be an invertible operator. Solving for $L[y(\chi)]$ leads to

$$L[y(\chi)] = -N[y(\chi)] \tag{9.D3}$$

Applying the inverse operator L on both sides of Eq. (9.D3) yields

$$y(\chi) = \phi(\chi) - L^{-1}\{N[y(\chi)]\} \tag{9.D4}$$

where $\phi(\chi)$ is a function that satisfies the condition $L[\phi(\chi)] = 0$. Now, assume that the solution y can be represented as infinite series of the form

$$\sum_{n=0}^{\infty} y_n(\chi) = \phi(\chi) - L^{-1}\left[\sum_{n=0}^{\infty} A_n(\chi)\right] \tag{9.D5}$$

where

$$\sum_{n=0}^{\infty} y_n(\chi) = y(\chi) \qquad A_n(\chi) = \frac{1}{n!}\left(\frac{d^n}{d\lambda^n} N\left\{\sum_{i=0}^{\infty}\left[\lambda^i y_i(\chi)\right]\right\}\right)_{\lambda=0} \qquad \sum_{n=0}^{\infty} A_n(x) = N[y(\chi)] \qquad n \geq 0$$

$$(9.D6)$$

Then, equating the terms in the linear system of Eq. (9.D5) gives the recurrent relation

$$y_0 = \phi(\chi) \qquad y_{n+1} = -L^{-1}(A_n) \qquad n \geq 0 \qquad (9.D7)$$

However, in practice, all the terms of series in Eq. (9.D.5) cannot be determined, and the solution is approximated by the truncated series $\sum_{n=0}^{N} y_n(\chi)$.

REFERENCES

Abdou M.A. and Soliman A.A. (2005) Variational iteration method for solving Burger's and coupled Burger's equations. *J. Comput. Appl. Math.* **181**, 245–251. DOI: 10.1016/j.cam.2004.11.032

Adomian G. (1984) Convergent series solution of nonlinear equations. *J. Comput. Appl. Math.* **11**, 225–230. DOI: 10.1016/0377-0427(84)90022-0

Adomian G. and Witten M. (1994) Computation of solutions to the generalized Michaelis-Menton equation. *Appl. Math. Lett.* **7**, 45–48. DOI: 10.1016/0893-9659(94)90009-4

Adomian G. (1995) Solving the mathematical models of neurosciences and medicine. *Math. Comput. Simul.* **40**, 107–114. DOI: 10.1016/0378-4754(95)00021-8

Amatore C., Oleinick A., Svir I., da Mota N., and Thouin L. (2006) Theoretical modeling and optimization of the detection performance: A new concept for electrochemical detection of proteins in microfluidic channels. *Nonlinear Anal. Model. Control* **11**, 345–365.

Angel Joy R., Meena A., Loghambal S., and Rajendran L. (2011) A two-parameter mathematical model for immobilized enzymes and homotopy analysis method. *Nat. Sci.* **3**, 556–565. DOI: 10.4236/ns.2011.37078

Anitha S., Subbiah A., Rajendran L., and Ashok Kumar J. (2010) Solutions of the coupled reaction and diffusion equations within polymer modified ultramicroelectrodes. *J. Phys. Chem. A* **114**, 7030–7037. DOI: 10.1021/jp1025224

Anitha S., Subbiah A., and Rajendran L. (2011a) Analytical expression of non-steady-state concentrations and current pertaining to compounds present in the enzyme membrane of biosensor. *J. Phys. Chem. A.* **115**, 4299–4306. DOI: 10.1021/jp200520s

Anitha S., Subbiah A., Subramaniam S., and Rajendran L. (2011b) Analytical solution of amperometric enzymatic reactions based on homotopy perturbation method. *Electrochim. Acta* **56**, 3345–3352. DOI: 10.1016/j.electacta.2011.01.014

Anitha A., Loghambal S., and Rajendran L. (2012) Analytical expressions for steady-state concentrations of substrate and product in an amperometric biosensor with the substrate inhibition—The Adomian decomposition method. *Am. J. Anal. Chem.* **3**, 495–502. DOI: 10.4236/ajac.2012.38066

Ariel P.D. (2010) Homotopy perturbation method and the natural convection flow of a third grade fluid through a circular tube. *Nonlinear Sci. Lett. A* **1**, 43–52.

Awawdeh F., Jaradat H.M., and Alsayyed O. (2009) Solving system of DAEs by homotopy analysis method, *Chaos Solitons Fractals* **42**, 1422–1427. DOI: 10.1016/j.chaos.2009.03.057.

Baronas R., Kulys J., and Ivanauskas F. (2006) Computational modelling of biosensors with perforated and selective membranes. *J. Math. Chem.* **39**(2), 345–362. DOI: 10.1007/s10910-005-9034-0

Baronas R. (2007) Numerical simulation of biochemical behaviour of biosensors with perforated membrane. In: Zelinka I., Oplatková Z., and Orsoni A. (eds.), *Proceedings 21st European Conference on Modelling and Simulation,* June 4th–6th, 2007, Prague, Czech Republic. DOI: 10.7148/2007-0214.

Baronas R., Ivanauskas F., and Kulys J. (2010) *Mathematical Modeling of Biosensors: An Introduction for Chemists and Mathematicians.* Springer Series on Chemical Sensors and Biosensors **9**, 104. DOI: 10.1007/978-90 481 3243 0

Biazar J. and Islam R. (2004) Solution of wave equation by Adomian's decomposition method and restrictions of the method. *Appl. Math. Comput.* **149**, 807–814. DOI: 10.1016/S0096-3003(03)00186-3

Biazar J. and Ghazvini H. (2009) He's homotopy perturbation method for solving systems of Volterra integral equations of the second kind. *Chaos Solitons Fractals* **39**, 770–777. DOI: 10.1016/j.chaos.2007.01.108

Bowden A.C. (2004) *Fundamentals of Enzyme Kinetics,* 3rd ed. Portland Press, London.

Cai X.C., Wu W.Y., and Li M.S. (2006) Approximate period solution for a kind of nonlinear oscillator by He's perturbation method. *Int. J. Nonlinear. Sci. Numer. Simulat.* **7**(1), 109–112. DOI: 10.1515/IJNSNS.2006.7.1.109

Carbanes J., Garcia-Canovas F., Lozano J.A., and Garcia-Carmona F. (1987) A kinetic study of the melanization pathway between L-tyrosine and dopachrome. *Biochim. Biophys. Acta—Gen. Subj.* **923**, 187–195. DOI: 10.1016/0304-4165(87)90003-1

Caruso F. and Schuler C. (2000) Enzyme multilayers on colloid particles: Assembly, stability, and enzymatic activity. *Langmuir* **16**, 9595–9603. DOI: 10.1021/la000942h

Caruso F., Fiedler H., and Haage K. (2000) Assembly of b-glucosidase multilayers on spherical colloidal particles and their use as active catalysts. *Colloids Surf. A* **169**, 287–293. DOI: 10.1016/S0927-7757(00)00443-X

Chowdhury M.S.H. and Hashim I. (2007) Solutions of time-dependent Emden–Fowler type equations by homotopy perturbation method. *Phys. Lett. A* **368**, 305–313. DOI: 10.1016/j.physleta.2007.04.020

Clark L.C. and Lyons C. (1962) Electrode systems for continuous monitoring in cardiovascular surgery. *Ann. N.Y. Acad. Sci.* **102**, 29–45. DOI: 10.1111/j.1749-6632.1962.tb13623.x

Coche-Guerante L., Labbe P., and Mengeand V. (2001) Amplification of amperometric biosensor responses by electrochemical substrate recycling. 3. Theoretical and experimental study of the phenol–polyphenol oxidase system immobilized in laponite hydrogels and layer-by-layer self-assembled structures. *Anal. Chem.* **73**, 3206–3218. DOI: 10.1021/ac0015341

Cook P.F. and Cleland W.W. (2007) *Enzyme Kinetics and Mechanism.* Garland Science, New York.

Decher G. and Hong J.D. (1991) Build up of ultrathin multilayer films by a self-assembly process: II. Consecutive adsorption of anionic and cationic bipolar amphiphiles and polyelectrolytes on charged surfaces. *Ber. Buns. Phys. Chem.,* **95**, 1430–1434. DOI: 10.1002/bbpc.19910951122

Domairry G. and Bararnia H. (2008) An approximation of the analytic solution of some nonlinear heat transfer equations: A survey by using homotopy analysis method. *Adv. Studies Theor. Phys.* **2**, 507–518.

Domairry G. and Fazeli M. (2009) Homotopy analysis method to determine the fin efficiency of convective straight fins with temperature-dependent thermal conductivity. *Commun. Nonlinear Sci. Numer. Simulat.* **14**, 489–499. DOI: 10.1016/j.cnsns.2007.09.007

Dong S. and Che G. (1991) Electrocatalysis at a microdisk electrode modified with a redox species. *J. Electroanal. Chem.* **309**, 103–114. DOI: 10.1016/0022-0728(91)87007-Q

Donnet J.B. and Basal R.C. (1984) *International Fiber Science and Technology, Carbon Fibers.* Marcel Dekker, New York.

Edmonds T.E. (1985) Electroanalytical applications of carbon fibre electrodes. *Anal. Chim. Acta* **175,** 1–22. DOI: 10.1016/S0003-2670(00)82713-0

Eswari A. and Rajendran L. (2010a) Analytical solution of steady state current at a microdisk biosensor. *J. Electroanal. Chem.* **641**, 35–44. DOI: 10.1016/j.jelechem.2010.01.015

Eswari A. and Rajendran L. (2010b) Analytical solution of steady-state current an enzyme-modified microcylinder electrodes, *J. Electroanal. Chem.* **648**, 36–46. DOI: 10.1016/j.jelechem.2010.07.002

Fang M., Grant P.S., McShane M.J., Sukhorukov G.B., Golub V.O., and Lvov Y. (2002) Magnetic bio/nanoreactor with multilayer shells of glucose oxidase and inorganic nanoparticles. *Langmuir* **18**, 6338–6344. DOI: 10.1021/la025731m

Forzani E.S., Solis V.M., and Calvo E.S. (2000) Electrochemical behaviour of polyphenol oxidase immobilized in self-assembled structures layer by layer with cationic polyallylamine. *Anal. Chem.* **72**, 5300–5307. DOI: 10.1021/ac0003798

Galceran J., Taylor S.L., and Bartlett P.N. (2001) Modelling the steady-state current at the inlaid disc microelectrode for homogeneous mediated enzyme catalysed reactions. *J. Electroanal. Chem.* **506**, 65–81. DOI: 10.1016/S0022-0728(01)00503-4

Ganji D.D. and Rafei M. (2006) Solitary wave solutions for a generalized Hirota–Satsuma coupled *KdV* equation by homotopy perturbation method. *Phys. Lett. A* **356**, 131–137. DOI: 10.1016/j.physleta.2006.03.039

Ghasemi M., Kajani M.T., and Babolian E. (2007) Numerical solutions of the nonlinear Volterra-Fredholm integral equations by using homotopy perturbation method. *Appl. Math. Comput.* **188**, 446–449. DOI: 10.1016/j.amc.2006.10.015

Ghori Q.K., Ahmed M., and Siddiqui A.M. (2007) Application of homotopy perturbation method to squeezing flow of a Newtonian fluid. *Int. J. Nonlinear Sci. Numer. Simulat.* **8**(2), 179–184. DOI: 10.1515/IJNSNS.2007.8.2.179

Golbabai A. and Keramati B. (2008) Modified homotopy perturbation method for solving Fredholm integro-differential equations. *Chaos Solitons Fractals*, **37**, 1528–1537. DOI: 10.1016/j.chaos.2006.10.037

Gonon F., Suaud-Changny M.F., and Buda M. (1992) In: Dittmar A. and Froment J.C. (eds.), *Proceedings of Satellite Symposium on Neuroscience and Technology*, Lyon, France.

Gue A.-M., Tap H., Gros P., and Maury F. (2002) A miniaturised silicon based enzymatic biosensor: Towards a generic structure and technology for multi-analytes assays. *Sens. Actuators B* **82**, 227–232. DOI: 10.1016/S0925-4005(01)01009-7

He J.H. (1998) Approximate analytical solution for seepage flow with fractional derivatives in porous media. *Comput. Meth. Appl. Mech. Eng.* **167**, 57–68. DOI: 10.1016/S0045-7825(98)00108-X

He J.H. (1999) Homotopy perturbation technique. *Comput. Math. Appl. Mech. Eng.* **178**, 257–262. DOI: 10.1016/S0045-7825(99)00018-3

He J.H. (2003a) A simple perturbation approach to Blasius equation. *Appl. Math. Comput.* **140**, 217–222. DOI: 10.1016/S0096-3003(02)00189-3

He J.H. (2003b) Homotopy perturbation method: A new nonlinear analytical technique. *Appl. Math. Comput.* **135**, 73–79. DOI: 10.1016/S0096-3003(01)00312-5

He J.H. (2005) Application of homotopy perturbation method to nonlinear wave equations. *Chaos Solitons Fractals* **26**, 695–700. DOI: 10.1016/j.chaos.2005.03.006

He J.H. (2006a) Homotopy perturbation method for solving boundary value problems. *Phys. Lett. A* **350**, 87–88. DOI: 10.1016/j.physleta.2005.10.005

He J.H. (2006b) Some asymptotic methods for strongly nonlinear equations. *Int. J. Mod. Phys. B* **20**(10), 1141–1199. DOI: 10.1142/S0217979206033796

He J.H. (2007) Variational iteration method—Some recent results and new interpretations. *J. Comput. Appl. Math.* **207**, 3–17. DOI: 10.1016/j.cam.2006.07.009

He J.H. and Wu X.H. (2007) Variational iteration method: New development and applications. *Comput. Math. Appl.* **54**, 881–894. DOI: 10.1016/j.camwa.2006.12.083

He J.H., Wu G.C., and Austin F. (2010) The variational iteration method which should be followed. *Nonlinear Sci. Lett. A* **1**(1), 1–30.

Hodak J., Etchenique R., Calvo E.J., Singhal K., and Bartlett P.N. (1997) Layer by layer self assembly of glucose oxidase with a poly(allylmanine)-ferrocene redox mediator. *Langmuir* **13**, 2708–2716. DOI:10.1021/la962014h

Ikeuchi H., Kawai Y., Oda Y., and Sato G.P. (1991) Determination of diffusion coefficients by means of normal pulse voltammetry with a stationary disk electrode. *Anal. Sci.* **7**, 389–392. DOI: 10.2116/analsci.7.389 fourteen

Indira K. and Rajendran L. (2011) Analytical expression of the concentration of substrates and product in phenol–polyphenol oxidase system immobilized in laponite hydrogels. Michaelis–Menten formalism in homogeneous medium. *Electrochim. Acta* **56**, 6411–6419. DOI: 10.1016/j.electacta.2011.05.012

Jafari H., Chun C., Seifi S., and Saeidy M. (2009) Analytical solution for nonlinear gas dynamic equation by homotopy analysis method. *Appl. Appl. Math.* **4**, 149–154.

Jaradat O.K. (2008) Adomian decomposition method for solving Abelian differential equations. *J. Appl. Sci.* **8**, 1962–1966. DOI: 10.3923/jas.2008.1962.1966

Junfeng L. (2007) Variational iteration method for solving two-point boundary value problems. *J. Comput. Appl. Math.* **207**, 92–95. DOI: 10.1016/j.cam.2006.07.014

Kissinger P.T. (2005) Biosensors—A perspective. *Biosens. Bioelectron.* **20**, 2512–2516. DOI: 10.1016/j.bios.2004.10.004.

Kulys J. and Baronas R. (2006) Modeling of amperometric biosensors in the case of substrate inhibition. *Sensors* **6**(11), 1513–1522. DOI: 10.3390/s6111513

Li S.J. and Liu Y.X. (2006) An improved approach to nonlinear dynamical system identification using PID neural networks. *Int. J. Non linear Sci. Numer. Simulat.* **7**(2), 177–182. DOI: 10.1515/IJNSNS.2006.7.2.177

Liao S.J. (1992) The proposed homotopy analysis technique for the solution of nonlinear problems. Ph.D. thesis, Shanghai Jiao Tong University, P.R. China.

Liao S.J. (2003) *Beyond Perturbation: Introduction to the Homotopy Analysis Method.* Chapman & Hall, CRC Press, Boca Raton, FL, 336.

Liao S.J. (2004) On the homotopy analysis method for nonlinear problems. *Appl. Math. Comput.* **147**, 499–513. DOI: 10.1016/S0096-3003(02)00790-7

Lvov Y. and Caruso F. (2001) Biocolloids with ordered urease multilayer shells as enzymatic reactors. *Anal. Chem.* **73**, 4212–4217. DOI: 10.1021/ac010118d

Lyons M.E.G., Bannon T., and Rebouillat S. (1998) Reaction/diffusion at conducting polymer ultramicroelectrodes. *Analyst* **123**, 1961–1966. DOI: 10.1039/A804039G

Manimozhi P., Subbiah A., and Rajendran L. (2010) Solution of steady-state substrate concentration in the action of biosensor response at mixed enzyme kinetics. *Sens. Actuators B* **147**, 290–297. DOI: 10.1016/j.snb.2010.03.008

McNaught A.D. and Wilkinson A. (1997) IUPAC. *Compendium of Chemical Terminology,* 2nd ed. (the "Gold Book"). Blackwell Scientific, Oxford, UK.

Meena A. and Rajendran L. (2010a) Analysis of pH-based potentiometric biosensor using homotopy perturbation method. *Chem. Eng. Technol.* **33**, 1–10.

Meena A., Eswari A., and Rajendran L. (2010b) Mathematical modelling of enzyme kinetics reaction mechanisms and analytical solutions of non-linear reaction equations. *J. Math. Chem.* **48**, 179–186. DOI: 10.1007/s10910-009-9659-5

Meena A. and Rajendran L. (2010c) Mathematical modeling of amperometric and potentiometric biosensors and system of non-linear equations—Homotopy perturbation approach. *J. Electroanal. Chem.* **644**, 50–59. DOI: 10.1016/j.jelechem.2010.03.027

Meena A., Eswari A., and Rajendran L. (2011) *In:* Pier Andrea Serra (ed.), *New Perspectives in Biosensors Technology and Applications.* In tech, Croatia, 215–228. DOI: 10.5772/936

Mohamed M.A. (2010) Comparison differential transformation technique with Adomian decomposition method for dispersive long-wave equations in (2+1)-dimensions. *Appl. Appl. Math.* **5**, 148–166.

Monosik R., Streansky M., and Sturdik E. (2012) Biosensors—Classification, characterization and new trends. *Acta Chim. Slovaca* **5**, 109–120. DOI: 10.2478/v10188-012-0017-z

Mousa M.M. and Ragab S.F. (2008) Application of the homotopy perturbation method to linear and nonlinear schrödinger equations. *Z. Naturforsch.* **63**(a), 140–144.

Nakamura H. and Karube I. (2003) Current research activity in biosensors. *Anal. Bioanal. Chem.* **377**, 446–468. DOI: 10.1007/s00216-003-1947-5

Odibat Z. and Momani S. (2007) A reliable treatment of homotopy perturbation method for Klein Gordon equations. *Phys. Lett. A,* **365**, 351--357. DOI: 10.1016/j.physleta.2007.01.064

Ozis T. and Yildirim A. (2007) A comparative study of He's homotopy-perturbation method for determining frequency-amplitude relation of a nonlinear oscillator with discontinuities. *Int. J. Nonlinear Sci. Numer. Simulat.* **8**(2), 243–248. DOI: 10.1515/IJNSNS.2007.8.2.243

Patel A. and Serrano S.E. (2011) Decomposition solution of multidimensional groundwater equations. *J. Hydrol.* **397**, 202–209. DOI: 10.1016/j.jhydrol.2010.11.032

Phanthong C. and Somasundrum M. (2003) The steady state current at a microdisk biosensor. *J. Electroanal. Chem.* **558**, 1–8. DOI: 10.1016/S0022-0728(03)00370-X

PonRani V.M.R.N., Rajendran L., and Eswaran R. (2011) Analytical expression of the substrate concentration in different part of particles with immobilized enzyme and substrate inhibition kinetics. *Anal. Bioanal. Electrochem.* **3**, 507–520.

Rahamathunissa G. and Rajendran L. (2008) Application of He's variational iteration method in nonlinear boundary value problems in enzyme–substrate reaction diffusion processes: Part 1. The steady-state amperometric response. *J. Math. Chem.* **44**, 849–861. DOI: 10.1007/s10910-007-9340-9

Ramana B.V. (2007) *Higher Engineering Mathematics.* Tata McGraw-Hill, New Delhi.

Revzin A.F., Sirkar K., Simonian A., and Pishko M.V. (2002) Glucose, lactate, and pyruvate biosensor arrays based on redox polymer/oxidoreductase nanocomposite thin-films deposited on photolithographically patterned gold microelectrodes. *Sens. Actuators B* **81**, 359–368. DOI: 10.1016/S0925-4005(01)00982-0

Rijiravanich P., Aoki K., Chen J., Surareungchai W., and Somasundrum M. (2006) Microcylinder biosensors for phenol and catechol based on layer-by-layer immobilization of tyrosinase on latex particles: Theory and experiment. *J. Electroanal. Chem.* **589**, 249–258. DOI: 10.1016/j.jelechem.2006.02.019

Scheller F. and Schubert F. (1992) *Biosensors.* Elsevier, Amsterdam.

Schulmeister T. and Pfeiffer D. (1993) Mathematical modelling of amperometric enzyme electrodes with perforated membranes. *Biosens. Bioelectron.* **8**, 75–79. DOI: 10.1016/0956-5663 (93)80055-T

Senthamarai R. and Rajendran L. (2010) System of coupled non-linear reaction diffusion processes at conducting polymer-modified ultramicroelectrodes. *Electrochim. Acta* **55**, 3223–3235. DOI: 10.1016/j.electacta.2010.01.013

Sevukaperumal S., Eswari A., and Rajendran L. (2011) Analytical expression pertaining to concentration of substrate and effectiveness factor for immobilized enzymes with reversible Michaelis-Menten kinetics. *Int. J. Comput. Appl.* **33**, 0975–8887. DOI: 10.5120/4004-5671

Shi G., Liu M., Zhu M., Zhou T., Chen J., Jin L., and Jin J.-Y. (2002) The study of Nafion/xanthine oxidase/Au colloid chemically modified biosensor and its application in the determination of hypoxanthine in myocardial cells in vivo. *Analyst* **127**, 396–400. DOI: 10.1039/B108462N

Siddiqui A.M., Hameed M., Siddiqui B.M., and Ghori Q.K. (2010) Use of Adomian decomposition method in the study of parallel plate flow of a third grade fluid, *Commun Nonlinear Sci. Numer. Simulat.* **15**, 2388–2399. DOI: 10.1016/j.cnsns.2009.05.073

Sohouli A.R., Famouri M., Kimiaeifar A., and Domairry G. (2010) Application of homotopy analysis method for natural convection of Darcian fluid about a vertical full cone embedded in porous media prescribed surface heat flux. *Commun. Nonlinear Sci. Numer. Simulat.* **15**, 1691–1699. DOI: 10.1016/j.cnsns.2009.07.015

Stamatin I., Berlic C., and Vaseashta A. (2006) On the computer-aided modelling of analyte-receptor interactions for an efficient sensor design. *Thin Solid Films* **495**, 312–315. DOI: 10.1016/j.tsf.2005.08.299

Sun H. and Hu N. (2004) Voltammetric studies of hemoglobin-coated polystyrene latex bead films on pyrolytic graphite electrodes. *Biophys. Chem.* **110**, 297–308. DOI: 10.1016/j.bpc.2004.03.005

Sweilam N.H. and Khader M.M. (2010) Approximate solutions to the nonlinear vibrations of multiwalled carbon nanotubes using Adomian decomposition method. *Appl. Math. Comput.* **217**, 495–505. DOI: 10.1016/j.amc.2010.05.082

Turner A.P.F., Karube I., and Wilson G.S. (1987) *Biosensors: Fundamentals and Applications.* Oxford University Press, Oxford, UK.

Varadharajan G. and Rajendran L. (2011a) Analytical solution of the concentration and current in the electroenzymatic processes involved in a PPO-rotating-disk-bioelectrode. *Nat. Sci.* **3**, 459–465. DOI: 10.4236/ns.2011.31001

Varadharajan G. and Rajendran L. (2011b) Analytical solutions of system of non-linear differential equations in the single-enzyme, single-substrate reaction with non-mechanism-based enzyme inactivation. *Appl. Math.* **2**, 1140–1147. DOI: 10.4236/am.2011.29158

Venugopal K., Eswari A., and Rajendran L. (2011) Mathematical model for steady state current at PPO-modified micro-cylinder biosensors. *J. Biomed. Sci. Eng.* **4**, 631–641. DOI: 10.4236/jbise.2011.49079

Wazwaz A.M. and Gorguis A. (2004) An analytic study of Fisher's equation by using Adomian decomposition method. *Appl. Math. Comput.* **154**, 609–620. DOI: 10.1016/S0096-3003(03)00738-0

Wollenberger U., Lisdat F., and Scheller F.W. (1997) *Frontiers in Biosensorics 2, Practical Applications*. Birkhauser Verlag, Basel.

INDEX

A

ab initio theory, 137
acetylcholinesterase, 257
activity, 5, 28, 31, 41, 49, 51, 65, 77, 83, 134, 156, 168, 169, 172, 191, 205, 211, 219, 222, 227, 238, 239, 243, 254, 257–259, 261, 262, 285, 316, 319, 329, 330, 332, 333, 339, 353, 354
activity coefficient, 168, 205, 243
Adomian decomposition method, 391
adsorption, 5, 6, 8–12, 15–17, 19, 20, 22–33, 61, 256, 268–271, 273, 278, 281, 297, 330
adsorption energy, 8–12, 15–17, 19, 22–27, 32, 33
advanced non-equilibrium models, 159, 164, 165
alternative current, 103
amorphous phase, 30
amperometric biosensor, 255, 260, 262, 264, 346, 348, 353, 363
  MNP-based, 262
amperometric immunosensors, 267
  AuNP-based, 267
amperometric sensor, 5, 42, 53, 54, 73, 255, 258, 277
amperometric type, 69, 72
amperometry, 252
analytical solution, 102, 118, 314, 348, 354, 356, 357, 361, 363, 364, 372, 374, 376, 385, 386, 391
anionic additives, 169
antibody, 252, 254, 268–277, 295, 340
antigen, 268–277
approximate analytical expression, 356
aptamer, 254
ascorbic acid detection, 331
atomic charge, 137, 143, 145, 147
AuNPs. *See* gold nanoparticles

B

back gate, 304, 305, 324, 325
bi-electrolyte structure, 76
binary electrolyte, 98
biocatalytic growth, 265, 266
biocomposite electrodes, 261
  CNT-based, 261
BioFET, 296
biosensor, 251–255, 257–267, 275–277, 280, 283–285, 296, 299, 307, 308, 310, 311, 321, 327–329, 331, 339–342, 344–346, 348, 353, 359, 362–364, 373, 374, 381–383, 385, 386
Bjerrum length, 312
Bode-type plots, 97, 116, 119, 121
bond length, 137, 145
boundary layer, 52, 96, 121, 122, 266
boundary-value problems, 346, 387
Buck-Stover equation, 166
bulk diffusion process, 55
bulk electroneutrality, 172
bulk of the solution, 203
Butler-Volmer equation, 51, 52, 75

C

calibration curve, 95, 103, 167, 181, 182, 185, 188, 189, 344, 363, 386
carbon dioxide sensor, 64
carbon nanotubes (CNTs), 29, 251–254, 258–262, 262, 265, 275, 280, 281, 284, 296–299, 332
carbon paste electrode, 257, 258, 264, 269, 270
catechol, 258, 350, 353, 375–377, 379–383
cation segregation, 19, 20
cation-exchange membrane, 97, 121
characteristic frequency, 114–116, 119, 122
charged associates, 172
charged interface, 202, 203
CHEMFET. *See* chemically selective field-effect transistor
chemical adsorption, 25, 27–30, 32
chemically selective field-effect transistor (CHEMFET), 201, 232, 236, 237, 242, 295
chemical potential, 6, 7, 25, 42, 43, 45–47, 81, 162, 204, 205, 218, 228
chemical reaction, 50, 59, 60, 76, 156, 164, 221, 331, 339
chiral selector, 137, 146, 148, 152
chronopotentiometric response, 101, 109, 123
CNT ISFET, 298, 299, 332
CNTs. *See* carbon nanotubes
co-exchanger, 173
co-extraction of the electrolyte, 168
cofactor, 256, 259
co-ion interference, 159, 168
colloidal gold-labeled antibody, 270
compartmental model, 215, 233, 234, 242
competitive equilibria, 134, 150
complexation, 133, 134, 145, 150, 169, 171, 173, 175, 179
composite solid electrolytes, 7, 29, 34
computer modeling, 155, 156, 158–160, 166, 169, 194
concentration profiles, 60, 111, 112, 180, 184–186, 189, 190, 193, 346, 347, 353, 354, 371, 372
conducting nanomaterials, 252
conductometric biosensor, 265, 266, 267
conductometric immunosensors, 276, 277
  AuNP-based, 276
  MNP-based, 277
continuity equation, 184, 188
CSAFM. *See* current-sensing AFM
current-sensing AFM (CSAFM), 298

D

Debye length, 123, 124, 184, 311–314
Debye screening length, 312, 322–324
Debye-Hückel theory, 205, 309
de-complexation, 133, 134, 150
denaturation, 259, 330
dendrimers, 265, 329
DEP. *See* dielectrophoresis
deprenyl, 138, 142
detection limit, 71, 72, 97, 158, 182, 185, 189, 190, 239, 257, 260, 267, 275, 276, 279, 283, 284, 322, 329
dielectric constant, 134, 136, 151, 205, 222, 308–311, 313, 319
dielectrophoresis (DEP), 298
differential capacitance, 319
diffuse charge layer, 319
diffuse layer, 9, 12, 17, 19, 27, 28, 222, 332
diffusion boundary layers, 96, 121, 122
diffusion coefficients, 55, 97, 99, 100, 109, 113, 116, 118, 119, 122–124, 156, 164, 165, 180, 186, 187, 190, 308, 314, 344, 345, 348, 350, 351, 362–365, 367–369, 375, 382, 383

diffusion layer model, 159, 179, 181
diffusion potential, 43, 161–164, 166, 170, 172, 174, 185, 188, 228
diffusion relaxation time, 110
dipole moment, 134, 136, 137, 145, 151
direct current, 114
direct enzyme wiring, 255, 258
displacement current equation, 183, 184
dissociation of the electrolytes in membranes, 170
divalent cations, 13, 172
DNA, 252, 254, 277–284, 295
DNA biosensor, 277, 280, 283, 284
  AuNP-based, 283
  CNT-based, 284
  MNP-based, 284
  nanomaterial-based, 277
DNA hybridization, 254, 278, 281, 284
Donnan equilibrium relations, 96, 97
Donnan potential, 101, 122, 227, 228
doping density, 303, 310, 312, 315
drift-diffusion equations, 307, 308
dual-gated nanowire ISFET, 296
dual-gate nanowire sensor, 304, 305, 324
dual-gate NW FET sensor, 324
dual-gate NW sensor. *See* dual-gate nanowire sensor
dual-gate ZnO ISFET, 326
dynamic range, 275, 331

E

effective capacitance model, 324
effectiveness factor, 384
electric circuit simulation program, 97, 121
electric current perturbations, 95, 101
electric potential, 17, 96, 97, 99–104, 106, 108–111, 121, 123, 124, 158, 272
electrified interface, 203
electrocatalytic activity, 261
electrochemical cell, 4, 73, 77, 204, 219
electrochemical CO_2 sensor, 45, 66–68
electrochemical DNA sensor, 280
  CNT-based, 280
electrochemical equilibrium, 158, 162, 166, 170, 171, 203, 208, 224
electrochemical impedance, 95–97, 101, 102, 109, 113, 117, 121, 170, 185, 273–276, 283, 284
electrochemical impedance spectroscopy, 95, 273, 276, 283, 284
electrochemical potential, 6, 42, 43, 45, 60, 162, 163, 204, 218, 226, 228
electrochemical sensors, 3–5, 41, 46, 53, 57, 76, 95–97, 121, 122, 155, 202, 252, 261, 271, 283
electrodeposition, 252, 257, 274
electrode potential, 42, 44, 50, 132, 156, 171, 180
electrodes, 3–6, 42–47, 50, 59, 60, 73–75, 82, 85, 95, 96, 103, 134, 135, 137–139, 141, 143, 144, 146, 148, 150, 151, 155–158, 160, 161, 168, 201, 206, 230, 252, 254–258, 260–262, 266, 267, 269, 275, 278, 281–284, 374, 376
  of the second kind, 44, 45
electrode transducer, 260
  CNT-coated, 260
electrodiffusion, 106, 107, 121, 315
  flow, 315
  process, 106
electrokinetic effect, 315
electrolyte, 3–7, 29, 34, 42–47, 49, 50, 58, 59, 65, 69, 76, 77, 86, 98, 100, 106, 167, 168, 170, 171, 173, 175, 182, 187, 201, 202, 204, 206, 207, 211, 219, 224, 228, 265, 295, 305, 307–309, 312–319, 321–324, 368, 375

electrolyte–oxide–semiconductor (EOS) capacitor, 206, 211, 224
electrolyte solution, 202, 204, 206, 207, 211, 219, 224, 228, 308
electromotive force, 156
electroneutrality, 121, 164, 172, 184, 187, 188
electron hopping rate, 253
electronic structure, 146, 147
electrophoretic force, 315
electrostatic potential, 137–139, 143, 144, 146, 225, 308, 312, 321
enantioanalysis, 137–139, 141, 142, 151, 152
enantiomer, 137
enantioselective, 137, 138, 141–144, 146, 148
energy diagrams, 23–26
energy minima, 137
ENFET, 329–331
enzymatic biosensor, 262–264
enzymatic reaction, 257, 265–267, 283, 341, 350, 351, 376, 386
enzyme, 252, 254–267, 269, 271, 272, 274, 280, 283, 295, 328–331, 339–341, 343, 344, 346, 348–351, 353, 354, 356–362, 364, 365, 368–370, 373–376, 382, 383, 385
enzyme biosensor, 258, 259
  CNT-based, 259
enzyme kinetics, 339, 341, 343, 353, 356, 357
enzyme sensor, 255
  nanomaterial-based, 255
enzyme–substrate complex, 354, 357
enzyme–substrate interaction, 359
enzyme–substrate reaction diffusion process, 359
EOS. *See* electrolyte–oxide–semiconductor (EOS) capacitor
equilibrium concentration, 10, 30, 31, 315
equilibrium potential, 51, 201
equilibrium potentiometric type, 57, 69
exchange current density, 51, 217
exhaust gas oxygen, 58
extraction, 133, 134, 150, 173

F

faradaic electric current, 99
Faraday constant, 57, 124, 156, 204, 217, 352
field-effect transistor, 201, 242, 295, 326
finite-difference method, 184, 188
first-order kinetics, 183, 381
fixed interference method, 239
FK model. *See* model of Frenkel and Kliever
fluidic environment, 307, 310
flux, 53–55, 57, 96–99, 106, 121, 122, 124, 163, 180, 183, 184, 187, 266, 365
fluxes of species, 163
fractal dimension of a surface, 315
Frenkel defects, 15, 32
fringing field, 311
fullerene, 138–149
fundamental frequencies, 137, 138

G

Galvani potential, 26, 42–45, 49, 132, 217, 219, 222, 226, 227
galvanic cell, 43, 46, 47, 62, 65, 82, 156
galvanostatic polarization, 189
gate functionalization, 328
genosensor, 279
glassy carbon electrode, 257, 259, 281, 283
glucose oxidase, 258, 259, 262, 263, 265–267, 329–331, 340

glucose-sensitive ENFET, 329
gold electrodes, 259, 268, 274–277, 284, 374
gold nanoparticles (AuNPs), 252, 253, 255–258, 262, 265–270, 273, 274, 276–279, 283, 284
Gouy-Chapman model, 9
Gouy-Chapman-Stern Model, 316
Gouy-Chapman theory, 222
group mobilities, 174

H

hardness, 137, 138, 145
Hartree-Fock theory, 137
Helmholtz layer, 9, 212, 332
Helmholtz layer capacitance, 332
Helmholtz-Perrin model, 215
Henderson equation, 185
heterogeneous surface rate constants, 183
high ionic conductivity, 4, 5, 7, 34
histidine, 138
homotopy analysis method, 383, 384, 388, 389
homotopy perturbation method, 349, 352, 364, 372–374, 376, 387
horseradish peroxidase (HRP), 258, 264, 268, 269, 271, 273, 274, 276
HRP. *See* horseradish peroxidase
hydration sheath, 211
hydrogen sensor, 77, 81, 82, 84, 85

I

immobilized enzyme, 274, 329, 376, 385
immunological FET, 295
immunomagnetic electrochemical sensors, 271
immunomagnetic impedancemetric sensors, 275
immunosensor, 255, 267–271, 273–277
  nanomaterial-based, 267
impedimetric biosensor, 265
impedimetric immunosensors, 274
  CNT-based, 274
impedimetric sensor, 273, 283
  AuNP-based, 273
initial (Cauchy) problem, 187
interdigitated electrode, 266, 267
interface defects, 3, 7, 8
interface interaction, 27, 28, 30
interface potentials, 43, 45
interfacial charge transfer, 161, 165, 168, 182
interfacial electrochemical equilibrium, 162, 171
intergrain boundary, 7, 27
intermolecular force, 137, 145
intrinsic buffer capacity, 319
ion diffusion, 161, 180
ion-exchange equilibrium, 168, 173
ion-exchange membrane, 95–98, 101, 103, 105–114, 116–118, 120, 121, 228
ion-exchange membrane systems, 95–98, 101, 109, 113, 117, 118, 121
ion-exchanger sites, 168, 170
ionic conduction, 41
ionic partition coefficients, 173, 183
ionic salt–oxide type, 28
ionic salts, 26, 31
ionic solids, 6
ionic strength, 205, 242, 309, 313–315, 324
ionic-strength adjuster, 206
ionic transport, 95–98, 106, 108, 121
ionic valency, 205, 229, 230
ionophore-based membranes, 155, 157–160, 232
ionophores, 157, 158, 162, 164, 168–177, 179, 187, 190

ion-pair complex, 133–136, 150
ion-selective electrode, 95, 96, 103, 155, 168
ion-sensitive field-effect transistor (ISFET), 201, 202, 206–211, 224, 225, 231–235, 237, 240–242, 295–301, 303–305, 326, 328–335
ion-sensitive glass electrodes, 201
ion-to-ionophore complexes, 170
iron oxide, 254, 262
ISE membrane, 157, 158, 160, 161, 164, 165, 169, 170, 172, 175, 184, 187, 189
ISE membrane modeling, 158, 160
ISFET. *See* ion-sensitive field-effect transistor
isoelectric point, 6, 14, 24, 28, 34
iteration methods, 169

J

junction potential, 43, 45

K

ketoprofene, 138
kinetic description of the ISE response, 181
Knudsen diffusion coefficient, 55
Knudsen diffusion process, 55, 56
Kreger-Vink diagrams, 31, 32

L

label-free detection, 275, 276, 281, 283
lactate oxidase, 331
Langmuir adsorption isotherm, 9
Laplace operator, 385
LAPS structure, 265
layer-by-layer (LbL) technique, 266
layer-by-layer self-assembly, 329
LbL technique. *See* layer-by-layer (LbL) technique
lean NO_x trap, 69
Lewis acid particles, 27
Lewis base centers, 27
ligand, 133, 134, 136, 137, 150–152, 157, 254, 273
limiting electric currents, 102
limiting-current-type sensor, 53, 54
liquid gate, 304, 324, 327
liquid membrane, 135–137
local equilibrium models, 179, 181, 182
lower detection limit of ISEs, 158
low-frequency impedance, 103–105, 114

M

macroscopic concepts, 155
magnetic core, 254
magnetic nanoparticles (MNPs), 252, 254, 255, 262, 267, 271, 276, 277, 284
material and charge balances, 169
mathematical modeling, 356, 359
McLean equation, 22
mechanism, 4–6, 8, 27, 28, 41, 131, 134, 137, 138, 145, 148–150, 158, 201, 203, 204, 211, 218, 224, 230, 232, 296, 315, 326, 331, 332, 335, 339, 348, 356, 357
mediators, 260
melting enthalpy, 30
melting point, 30, 71
membrane, 72, 95–98, 100–103, 105–114, 116–118, 120–123, 131–138, 141, 143, 144, 146, 148–151, 155, 157–166, 168–172, 174, 175, 177–194, 208, 209, 226–236, 242, 268, 269, 273, 274, 295, 328, 332, 334, 335, 340, 344, 363, 364
membrane-based ISFET, 208, 225, 233

membrane configuration, 132, 133
membrane diffusion potential, 188, 228
membrane electric potential, 97, 101, 111, 121
membrane equilibria, 131, 133, 149–151
membrane ISFETs (MEMFETs), 295
membrane–oxide interface, 209
membrane resistance, 97, 122, 123
membrane–solution interface, 131–136, 149–151, 162, 166, 168, 170, 171, 180, 181, 183, 209, 228
membrane-solution interface, 134
membrane-through transference number, 165
MEMFETs. *See* membrane ISFETs
metallic nanoparticles, 252, 265
metal–solution junction, 211–213, 216, 217, 219, 221
method of lines, 185
Michaelis-Menten kinetics, 341, 343–346, 348, 349, 353, 364, 375, 376, 382, 386
microcylinder biosensors, 373, 382
microdisk biosensors, 363, 364
microelectrode, 160, 161, 269, 274, 373
micro ISFET, 335
migration, 104, 171, 184
mixed enzyme kinetics, 353
mixed-ion solution, 237
mixed potential, 50–52, 60, 69, 70, 73
mixed potentiometric type, 73
MnO_2 nanoparticles, 330, 331
MNPs. *See* magnetic nanoparticles
mobilities of ions, 163, 180
model of Frenkel and Kliever (FK model), 6, 14, 15
molecular interaction, 131, 137, 151
  modeling of, 131, 151
Morf-Bakker selectivity coefficient, 177
MOS capacitor, 206
Mott-Schottky model, 7, 9
multispecies approximation, 159, 170–172, 174, 175, 194
multispecies model, 170, 171, 174
multiwall carbon nanotubes (MWCNTs), 253, 260, 261, 281
MWCNTs. *See* multiwall carbon nanotubes

N

NaCl, 8, 10, 13, 14, 18, 109, 113
NaCl crystals, 13
NaCl doped with bivalent impurities, 14
NaCl-type crystals, 8, 10, 13, 18
NADH, 253, 254, 261
Nair-Alam model, 296, 307
nanobiocomposite film, 329
nanocantilever, 300, 301
nanocomposite solid electrolytes, 29
nano-ISFET, 296, 298, 301, 316, 321, 334
nanomaterial, 251, 252, 254, 262, 284, 332
nanoparticle, 29, 251–255, 257, 258, 260, 262, 263, 265–271, 273, 274, 276–280, 283, 284, 296, 328–331, 334
  role of, 328–331
nanoparticle-modified gate, 295
nanoporous silicon ISFET, 297
nanoscale ISFET, 297
nanostructured channel, 295, 328
nanowire diameter, 303, 310, 316, 321
  effect of, 310
nanowire length, 310
  effect of, 310
nanowire (NW) sensor, 304, 305, 312, 315, 316, 322, 324
  energy band model of, 305
Nernst equation, 47, 82, 232
Nernstian behavior, 103, 184

Nernstian relation, 46
Nernstian response, 157, 219, 222, 230, 326
Nernst limit, 324, 325
Nernst-Planck equation, 163, 187
Nernst-Planck flux equations, 96–98, 121
network simulation method, 96, 97, 121
neuron model, 332
neutral associates, 170
Newton-Raphson method, 184
Nikolsky-Eisenman equation, 103, 157, 166, 167
Nikolsky-Eisenman formalism, 228, 233, 242
Nikolsky-Eisenman equation, 103
Nikolsky equation, 156, 157, 175, 181, 230, 232, 233, 241
non–Michaelis-Menten kinetics, 341, 343, 353
non-Nernstian response slopes, 159, 168
nonautonomous phase, 30
nonconducting nanomaterial, 252, 254
nonequilibrium modeling, 181
  in real time and space, 181
nonlinear equations, 341, 351, 386–388
nonlinear phenomena, 341
nonlinear reaction diffusion equations, 341, 349, 372, 386
non-uniform grid, 185
NO_x sensor, 67–69, 71–73, 75
NW operational modes, 311
  effect of, 311
Nyquist spectra, 276
Nyquist-type plot, 113, 115

O

objective function method, 173
Ohmic resistance, 103–105, 110, 115, 116, 119, 122, 332
one-dimensional approach to ISE membrane modeling, 160
open-circuit potential, 44
open-circuit voltage, 44, 82
optimal sensor performance, 96, 122
ordinary differential equations, 187, 357, 359
overall membrane potential, 160–162, 164, 185, 187, 190, 194
oxidase, 256–259, 262, 263, 265–267, 329–331, 340, 349, 352, 382
oxidation, 216, 217, 254, 256, 264, 269, 277, 279, 281, 331, 363
oxide–solution junction, 206, 219
oxygen sensor, 45, 49, 57–63, 85

P

partial differential equations, 184, 187, 340, 347, 373
perovskite oxides, 78
phase boundary model, 158, 164, 166, 170, 185
phase boundary theory, 166
pH ISFET, 207–209, 224, 232, 234
pH response of silicon nanowires, 316
physical models, 155, 156, 158, 160, 161
planar ISFET, 297, 304, 328, 332
plasticizer, 134, 136, 151, 171
plastic membrane, 136, 137, 151
point defects, 4, 6–8, 15, 25, 27, 34, 80
Poisson-Boltzmann equation, 307, 309
Poisson equation, 184, 187, 307, 308
polarity, 136, 151, 170, 171, 311, 312, 317
polymeric membranes, 157, 162
polyphenol oxidase, 349, 352, 382
potential development, 131–134, 137, 138, 148, 150

potential difference, 26, 42–44, 46, 49, 50, 101, 106, 108, 109, 122, 160, 162, 163, 201, 203, 204, 217, 219, 221, 222, 225, 226
potentiometric biosensors, 345
potentiometric cell, 103
potentiometric chemical sensor, 202
potentiometric selectivity, 157, 229
potentiometric sensor, 3, 5, 46, 47, 49, 53, 64, 79, 85, 131–134, 136–138, 142, 149–151, 155, 189, 265, 272
  AuNP-based, 272
potentiometry, 201, 252
PPO-modified microcylinder biosensors, 373, 382
protective shells, 254
PSpice, 97, 108, 113, 118, 121

Q

quinone, 264, 375, 381, 382

R

rate constant, 156, 183, 186, 189, 259, 346, 348, 355, 358, 359, 375
reaction diffusion equations, 341, 349, 372, 386
real space and time domains, 187
reduction, 52, 67, 68, 72, 216, 217, 255, 258, 284, 304, 315, 327, 363, 366, 373
re-extraction, 133, 134
resistivity sensors, 30
response of the electrode, 134

S

salt diffusion coefficient, 109, 113, 122
SAM. *See* self-assembled monolayer
Schottky defects, 10, 12, 13, 24, 26, 32
screening effect, 264, 312, 315
screen-printed electrode, 261
segmented sandwich membrane potential, 179
segregation energy, 20, 22–24, 27
selective adsorption of cations, 27
selective catalytic reduction, 68
selectivity, 5, 6, 33, 95, 131, 134, 149–151, 156–158, 160, 167, 171, 174–177, 181, 182, 185, 208, 228–230, 232, 234, 238, 242, 264, 328, 329
selectivity coefficient, 156, 167, 174–177, 181, 229, 238, 242
self-assembled monolayer (SAM), 257, 269, 273, 276, 279, 326
semiconducting and spinel-type oxide electrodes, 73
semiconductor, 27, 32, 201, 206, 211, 224–226, 265, 296, 299, 308, 313, 328
sensing chemical species, 41
sensor response, 66, 81, 97, 134, 137, 158, 239, 374, 377–379
separation solution method, 239
short-circuit potentiometric type, 75
Sieverts' law, 83
silicon nanowire ISFET, 298, 305
simulation, 95–97, 108, 113, 114, 118, 120–122, 156, 159, 166, 168–171, 175, 177, 178, 181, 184, 188, 189, 191, 194, 309, 310, 315, 332, 333, 339, 344, 345, 354, 386
single-gate nanowire sensor, 304, 305
single-gate sensors, 304
single-gate structure, 304
single-wall carbon nanotubes (SWCNTs), 253, 258–260, 332, 333
Si-NW biosensor, 299, 307
Si-NW ISFET, 299, 303, 304
SiO_2 nanoparticle, 329
site-binding model, 222
site filling factor, 179, 180

size effects, 5, 8, 25, 29
slope, 16, 52, 64, 66, 118, 134–137, 140, 141, 143, 146, 148, 151, 169, 175, 181, 231, 238, 242, 276, 297, 344
small-signal ac response, 121
solid electrolytes, 3–5, 7, 29, 34, 45, 69
solid-phase spreading, 30
solid-state electrochemical, 3, 41, 202
solid-state reference electrode, 83
solid-state sensor, 203
solution–insulator interface, 207
SO_x sensor, 76
space charge, 5–9, 27, 300
space-charge region, 164
spherical biosensor, 383, 385
stability constant, 134, 135, 150, 151, 168, 175
stability problems, 76
steady state, 54, 55, 102, 108, 112, 170, 181, 185–187, 346, 351, 355, 364, 375, 382
Stern capacitance, 319
Stern layer, 319
Stern model, 6–9, 12, 14–16, 20, 24–26, 34
subthreshold regime, 321–323, 328
superionic oxide ceramics, 15
superionic oxides, 7, 8, 15, 19, 23, 24
surface conductivity, 8, 33
surface defects, 6, 8, 9, 31, 33
surface disorder, 23
surface potential, 6–24, 26, 27, 30, 32–34, 225, 300, 319, 322
surface potential in silver halides, 15
SWCNT model, 332
SWCNTs. *See* single-wall carbon nanotubes

T

Teorell-Meyer-Sievers model, 227, 228
TFET. *See* tunnel field-effect transistor
Thiele modulus, 385
three-way catalyst, 57
time-dependent electrode potential, 180
total energy, 137, 140, 145
total equilibrium model, 157, 170
total potential for the cell, 45
transference number, 44, 165
  of IzI species, 165
transport number, 81, 99, 109, 113, 123
trienzyme biosensor, 362
tunnel field-effect transistor (TFET), 326, 328
tyrosinase, 257, 258, 263

U

urease, 262, 266, 267

V

variational iteration method, 346, 347, 357, 360, 386, 387
voltammetry, 270, 271, 282
Volta potential, 225

W

Warburg factor, 123
Warburg impedance, 123
waveform, 97, 116, 117, 121

Y

yttria-stabilized zirconia, 3, 4, 50, 58

Z

zero-order catalytic kinetics, 370
zirconia electrolyte, 58, 59
zirconia oxygen sensor, 60, 61